Business and Climate Change Governance

Governance and Limited Statehood Series

Series editors:
Thomas Risse, Professor of International Politics, Freie Universität Berlin, Germany

Ursula Lehmkuhl, Professor of Modern History, Freie Universität Berlin, Germany

This ground-breaking monograph series showcases cutting edge research on the transformation of governance in countries with weak state institutions. Combining theoretically-informed and empirically grounded scholarship, it challenges the conventional governance discourse which is biased towards modern developed nation-states. Instead, the series focuses on governance in Africa, Asia and Latin America including transnational and transregional dimensions.

Located at the intersection of global governance and international relations, on the one hand, and comparative politics, area studies, international law, history, and development studies, on the other, this innovative series helps to challenge fundamental assumptions about governance in the social sciences.

Governance and Limited Statehood Series
Series Standing Order ISBN 978–0–230–23597–7 (hardback) and
ISBN 978–0–230–23598–4 (paperback)

You can receive future titles in this series as they are published by placing a standing order. Please contact your bookseller or, in case of difficulty, write to us at the address below with your name and address, the title of the series and the ISBNs quoted above.

Customer Services Department, Macmillan Distribution Ltd, Houndmills, Basingstoke, Hampshire RG21 6XS, England

---

Business and Climate Change Governance

South Africa in Comparative Perspective

Edited by

Tanja A. Börzel

Professor, Otto-Suhr-Institute for Political Science, Freie Universität Berlin, Germany

and

Ralph Hamann

Associate Professor, Graduate School of Business, University of Cape Town, South Africa

palgrave
macmillan

First published 2013 by
PALGRAVE MACMILLAN

Palgrave Macmillan in the UK is an imprint of Macmillan Publishers Limited, registered in England, company number 785998, of Houndmills, Basingstoke, Hampshire RG21 6XS.

Palgrave Macmillan in the US is a division of St Martin's Press LLC, 175 Fifth Avenue, New York, NY 10010.

Palgrave Macmillan is the global academic imprint of the above companies and has companies and representatives throughout the world.

Palgrave® and Macmillan® are registered trademarks in the United States, the United Kingdom, Europe and other countries.

ISBN 978–1–137–30273–1

This book is printed on paper suitable for recycling and made from fully managed and sustained forest sources. Logging, pulping and manufacturing processes are expected to conform to the environmental regulations of the country of origin.

A catalogue record for this book is available from the British Library.

A catalog record for this book is available from the Library of Congress.

Typeset by MPS Limited, Chennai, India.

Contents

List of Tables

List of Figures

Acknowledgements

This book is the outcome of a collaborative project involving research-ers at the University of Cape Town (UCT), South Africa, and the Collaborative Research Centre 700 on Governance in Areas of Limited Statehood, which is headquartered at the Freie Universität Berlin in Germany. The project was supported by the South African National Research Foundation and the German Bundesministerium für Bildung und Forschung. The South African researchers were also supported by the UCT Vice-Chancellor's Strategic Fund and the UCT African Climate and Development Initiative. The German researchers benefitted from funding by the German Research Foundation (DFG).

Our special thanks go to Jana Hönke, Thomas Risse, Clifford Shearing, and Gina Ziervogel for vital contributions throughout the project, as well as Christine Colvin, Andre Fourie, Jon Hanks, Mark New, Steve Nicholls, Dirk Visser, Martine Visser, Tatjana von Bormann, and other participants in a public workshop held at the UCT Graduate School of Business in April 2012. We are also grateful to Lisa van Hoof, Ruben Kremers, and Klara Schwobe for the fantastic job they did on formatting and editing this book.

Tanja A. Börzel, Berlin, and Ralph Hamann,
Cape Town

Notes on Contributors

Editors

Tanja A. Börzel is Professor of Political Science and Chair of European integration at the Freie Universität Berlin, Germany. She co-directs the Research College 'The Transformative Power of Europe' with Thomas Risse and conducts several research projects on governance in areas of limited statehood.

Ralph Hamann is Research Director and Associate Professor at the Graduate School of Business, University of Cape Town. His research is on organisational and governance responses to complex social and environmental problems, including climate change, food security, and mining company–community relations. He is also an Extraordinary Associate Professor at the School of Public Management and Planning, University of Stellenbosch, Chair of the Southern Africa Food Lab, and a Director of the Cape Town Partnership.

Contributors

John Fay has a PhD from the Graduate School of Business at the University of Cape Town. His research focus is on incentive approaches to address climate change in Africa. He holds a Bachelor's degree in History from Duke University and an MBA from the Johnson School at Cornell University.

Jan Froestad is Associate Professor at the Department of Administration and Organisation Theory, University of Bergen, Norway. He was appointed as an Honorary Research Associate, University of Cape Town, in July 2011.

Tom Herbstein is a PhD student and Senior Research Coordinator at the University of Cape Town's Centre of Criminology. He is involved in the management of the Centre's 'Fulcrum Institution Programme', which explores the capacity of large regulatory institutions to help engage in the management of climate change.

Christopher Kaan is a PhD candidate at the Freie Universität Berlin. His research focuses on private business regulation and corporate social responsibility in transnational supply chains.

Farai Kapfudzaruwa is a researcher and lecturer in the Old Mutual Emerging Markets Research Group at the Graduate School of Business (GSB), University of Cape Town. He recently completed his PhD at the GSB with a thesis focusing on business responses to climate change in areas of limited statehood. His other research interests include inclusive business and corporate sustainability.

Stine Klapper was a Research Associate at the Center for Transnational Relations, Foreign and Security Policy at the Freie Universität Berlin, Germany, working for the project 'Sustainable Development Reflexive Inputs to World Organisation' until October 2012. She now works for the Friedrich-Ebert-Stiftung in Prague, Czech Republic.

Nicole Kranz is Policy Advisor in international water policy and infrastructure at German International Cooperation (GIZ). She completed her PhD in political science at the Collaborative Research Center 'Governance in Areas of Limited Statehood' at the Freie Universität Berlin focusing on water management practices of mining companies in South Africa.

Nadine Methner is a PhD candidate at the Department of Environmental and Geographical Science at the University of Cape Town. Her research interests are in adaptive and sustainable water resource management and the potential for collective action and social learning in complex socio-ecological systems.

Deon Nel is the head of WWF South Africa's Biodiversity Unit, where he has developed a number of successful multi-stakeholder partnerships with the business sector. He serves on WWF's international conservation committee, which guides the organisation's global conservation strategy. Deon holds a PhD in Ecology and an MPhil in Environmental Law, and is a research associate with the University of Cape Town's Department of Public Law.

Moliehi Shale is a PhD student in the Centre of Criminology at the University of Cape Town. Her dissertation explores the regulation of risk in the informal insurance sector. Shale is a fellow of the Alexander von Humboldt Stiftung International Climate Protection Fellowship 2012/3 and is hosted by the SFB700 at Freie Universität, Berlin.

Clifford Shearing holds the South African National Research Foundation Research Chair in Security and Justice at the University of Cape Town, where he also holds the Chair of Criminology and directs the Law Faculty's Centre of Criminology.

Christian R. Thauer is a Lecturer in International Relations at the Center for Transnational Relations, Foreign and Security Policy at the Freie University Berlin. He holds a PhD from the European University Institute in Florence. In his thesis, he analysed internal drivers for businesses' social conduct. In the past four years, he also worked for the collaborative Research Center 'Governance in Areas of Limited Statehood'.

List of Abbreviations

ABI	Association of British Insurers
AGOA	African Growth and Opportunity Act
BRICS	Brazil, Russia, India, China and South Africa
BUSA	Business Unity South Africa
CAPM	Capital Asset Pricing Model
CDM	Clean Development Mechanism
CPI	Consumer Price Index
CSR	Corporate Social Responsibility
DJSI	Dow Jones Sustainability Index
EEA	Energy Efficiency Accord
EIUG	Energy Intensive User Group
FDI	Foreign Direct Investment
FfF	Farming for the Future
FIT	Feed in Tariff
GBJ	Good Business Journey
GEAR	Growth, Employment and Redistribution
GHG	Greenhouse Gas Emissions
IEA	International Energy Agency
IPP	Independent Power Producer
IPCC	Intergovernmental Panel on Climate Change
IRR	Internal Rate of Return
JSE	Johannesburg Stock Exchange
KAM	Kenya Association of Manufacturers
Kwh	Kilowatt Hour
MAP	Market Incentive Programme
MBI	Market-Based Incentive
MWh	Megawatt hour
NAS	National Adaptation Strategy

NBI	National Business Initiative
NCCRS	The National Climate Change Response Strategy
NHTSA	National Highway Traffic Safety Administration
NSE	Nairobi Stock Exchange
NERSA	National Energy Regulator of South Africa
NGO	Non-Governmental Organisation
OECD	Organisation for Economic Co-operation and Development
ROSCA	Rotating Savings and Credit Associations
SARI	South African Renewables Initiative
UCT	University of Cape Town
UN	United Nations
UNDP	United Nations Development Programme
UNEP-FI	United Nations Environmental Programme – Finance Initiative
UNFCCC	United Nations Framework Convention on Climate Change
WWF	World Wide Fund for Nature
YTM	Yield to Maturity

1
Business Contributions to Climate Change Governance in Areas of Limited Statehood: Introduction

Ralph Hamann and Tanja A. Börzel

Introducing our research question

Climate change and related social and environmental changes present societies with increasingly urgent problems at multiple spatial scales. A diverse range of social actors is involved in contributing to climate change, and an even broader range is suffering and will suffer its manifold consequences. In this book we focus on business organisations because they are vital arenas of decision-making, in both mitigation and adaptation. In other words, they are important sources of greenhouse gas emissions (GHG), and they therefore need to help to mitigate these emissions. At the same time, these businesses also need to make significant changes to prepare for and adapt to increasingly severe climate change and the resulting socio-economic changes (Berkhout et al., 2006; Kolk and Pinkse, 2012).

We ask how and why business organisations contribute to climate change governance. Governance is broadly about 'organised efforts to manage the course of events in a social system' (Burris et al., 2008: 3). We define it here more specifically as the structures and processes through which commonly binding rules are established, and through which public goods or services are provided (Risse, 2011). Climate change governance, then, pertains first to those efforts that seek to maintain a stable climate as an essential, global public good – including the norms, rules and procedures that aim to mitigate climate change through reduction of atmospheric greenhouse gases, through either reduced emissions or sequestration. Second, climate change governance seeks to ensure that social and economic systems are able to adapt, and effectively respond, to climate changes that are already occurring and that will occur in future. This is commonly referred to as climate change adaptation (IPCC, 2007). To some extent, mitigation and adaptation are distinctive policy arenas, because 'their impact refers to different temporal and spatial scales [... and they] involve different actors in the process of policy formulation and implementation' (Pinkse and Kolk, 2012: 181). But in some areas, notably land use change and forests, there are important synergies between adaptation and mitigation (e.g. Swart et al., 2003).

The focus on business contributions to climate change governance links this book to a broader debate about the role of private actors in governance, and one of our objectives is to synthesise the partially separate manifestations of these themes in the political science and management studies literatures. The role of private actors in governance is still contested. In an increasingly globalised economy, companies are assumed to escape strict national regulation by relocating their production sites to 'areas of limited statehood' where state regulation is low and/or enforcement is weak. While countries with high levels of regulation will respond by lowering their standards, countries with weak regulatory capacities refrain from tightening regulation in order not to lose foreign direct investment. Thus, the behaviour of firms can drive states into a 'race to the bottom', leading to degradation of natural resources and to compromising social standards for the sake of potential economic growth or attracting short-term foreign investment (see, e.g. Chan and Ross, 2003; Brühl et al., 2001; Kaufmann and Segura-Ubiergo, 2001; Lofdahl, 2002). At the same time, however, companies can be 'drawn into playing public roles to compensate for governance gaps and governance failures at global and national levels' (Ruggie, 2004a: 13). Empirical evidence suggests that under some conditions companies voluntarily commit themselves to social and environmental standards and adopt private self-regulatory regimes – even in the absence of a regulatory threat by the state (e.g. Mol, 2001; Vogel and Kagan, 2004; Levy and Newell, 2002; Pinkse and Kolk, 2009; Prakash and Potoski, 2007; Greenhill et al., 2009; Börzel and Thauer, 2013). Climate change is a formidable research area in which to investigate whether business performs such governance functions, what forms these contributions take, and under what conditions and for which motives and incentives business performs them.

This introductory chapter provides an initial review of the literature and a conceptual framework for the book as a whole. In the next section, we explain in more detail what we consider to be business contributions to climate change governance. We then discuss the issue of climate change as a 'wicked problem' in order to describe the problem characteristics and organisational field of interest in our empirical analyses. We also explain our focus on 'areas of limited statehood', defined as geographical or thematic areas in which states struggle or even fail to establish and enforce commonly binding rules and to provide or safeguard public goods. The concept of limited statehood helps us first, to investigate climate change as an organisational field in which states face fundamental challenges, and second, to analyse business contributions to governance in particular geographical areas, in which states face capacity constraints.

In the fourth section, we develop a framework with which to characterise firm responses to climate change and to address the first question of this book: *how* do companies contribute to climate change governance? We develop this framework by combining existing typologies in either climate change mitigation or adaptation, and use it to provide an overview of the business activities discussed in the chapters of this book. Finally, in the fifth section, we consider a set of concepts to help answer the second question of this book: *why* do companies contribute to climate change governance, and what explains the variance in firms' responses? These concepts fit broadly into three categories – institutional drivers, organisational drivers and problem characteristics – and their relative importance and potential interaction will be prominent themes in the empirical analyses and concluding chapter of this book.

Business contributions to climate change governance

Almost by definition, much of the governance literature emphasises the role of either public or private regulation in curbing the negative externalities associated with private organisations' activities. This is a vital concern, given the growing emissions of greenhouse gases, most of which are attributable to business organisations. Climate change that is brought about as an external consequence of emissions from the combustion of fossil fuels and fugitive emissions from smokestacks, exhausts, refrigerators and various other business processes has thus been framed as a large-scale market failure by prominent government advisor, Nicholas Stern:

> The problem of climate change involves a fundamental failure of markets: those who damage others by emitting greenhouse gases generally do not pay ... Climate change is a result of the greatest market failure the world has seen.[1]

The market failure is at the same time a governance failure, due to the absence of binding rules and ongoing difficulties in establishing and enforcing these rules in the international climate change regime. This is why establishing and enforcing binding rules is a vital part of our definition of climate change governance, and why we want to understand better how and why companies respond to such rules and sometimes even contribute to their development, even if states are reluctant or incapable to do so.

At the same time, companies are increasingly expected not just to adopt commonly binding rules, but also to develop innovative business processes, products and services that help society reduce emissions and/or adapt to climate change. Such contributions to the delivery of public goods associated with climate stability or social resilience may be motivated by the need to build companies' competitive advantage and to establish and maintain a conducive business context (Porter and Kramer, 2011; Börzel and Thauer, 2013). Indeed, such hopes and expectations are expressed even at the highest levels of the United Nations climate change machinery, as illustrated in the following comments made by Christiana Figueres (Executive Secretary of the UN Framework Convention on Climate Change) to a meeting of business leaders:

> Business has the power to change consumer and supplier behaviour and turn it into a powerful vocal support that gives policy makers a clearer space in which to act ... Admittedly, in the absence of a clear solid international framework, this represents a risk for business, but I pose that it is a manageable risk ... Your [business] leadership is essential: 1. To push the envelope within your own business; 2. To bring around others in your business field to be ambitious; and 3. To create a virtuous cycle of push and pull between public and private sectors to pave the road toward sustainability and a low carbon future.[2]

This 'push and pull between public and private sectors', illustrated in Figure 1.1, is a prominent theme in this book. Climate change has created an issue-specific organisational field, a platform for 'debates in which competing interests negotiate over issue interpretation' (Hoffman, 1999: 353), in which institutions create a set of expectations to be fulfilled or options from which to choose (Hoffman, 2001). Climate change has led to widespread and diverse changes in regulative, normative and cognitive institutions, and these exert pressure on companies to respond. For example, while state-based rules on climate change in Africa are as yet relatively limited, initiatives such as the Global Reporting Initiative or the Carbon Disclosure Project have created an important set of normative and cognitive structures.[3] We thus expect a variety of institutional factors to drive firms to engage in activities related to climate change mitigation or adaptation. These institutional drivers are complemented or enhanced by firm-internal drivers, such as expected efficiency gains or increased market share, or benefits for a firm's reputation. Organisations are not just passive respondents to institutional forces, but have diverse interests and display some urgency in responding within their context (DiMaggio, 1991; Perrow, 1985). Companies' characteristics and strategic orientation are likely to influence how they respond to institutional pressures or incentives. In particular, management scholars have highlighted the diverse ways in which firms may seek to enhance their competitiveness in response to environmental concerns. This interplay between companies and their institutional context in influencing firms' responses – the top arrow in Figure 1.1 – is discussed in more detail in a dedicated section below.

Companies' responses will inadvertently, or even purposefully, influence the organisational field in which they are active – that is, there is likely to be some influence such as that represented by the bottom arrow in Figure 1.1. Some organisations may make a strategic choice to become institutional entrepreneurs – 'actors who have an interest in particular institutional arrangements and who leverage resources to create new institutions or to transform existing ones' (Maguire et al., 2004: 657). The organisational field surrounding climate change is an emerging one, especially in areas of limited statehood. Given, also, the significant business interests at stake in the climate change policy arena (such as the imposition of carbon taxes, for example), there are important incentives for companies to try to influence the organisational field

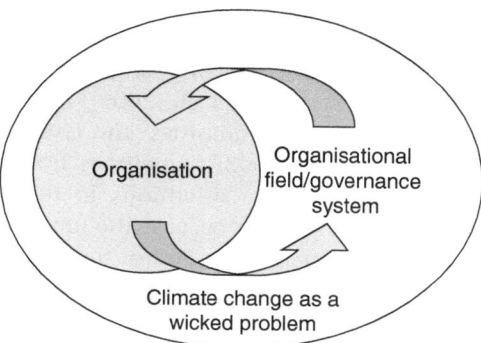

Figure 1.1 Schematic representation of the interactions between organisations and the governance system

proactively. This is likely to include lobbying to achieve particularistic interests, which of course has already received significant scholarly attention (Ikwue and Skea, 1994; Kolk, 2001; Levy, 1997, Newell and Paterson, 1998). According to some analyses, such overtly confrontational approaches have shifted toward more conciliatory or even cooperative forms of engagement with the public sector (Kolk, 2001; Levy and Kolk, 2002; Levy and Newell, 2002; Levy and Rothenberg, 2002). The bottom arrow in Figure 1.1 may even include proactive efforts by firms to raise anti-pollution standards in public policy, or support the implementation of such regulations, for reasons discussed in more detail below.

We take a broad approach to climate change governance that includes activities within firm boundaries, as long as they are intentional and have some benefits for climate change mitigation and/or adaptation. In the mitigation arena, this 'intentionality' is relatively clear, especially if companies seek to participate in voluntary or formal carbon markets, as greenhouse gas reductions then need to fulfil the Clean Development Mechanism's requirement of 'additionality' (that is, the reductions in question would not have been affected regardless). For other mitigation efforts, such as energy efficiency measures motivated in the first instance by cost savings, this intentionality can be less obvious.

The link between intention and action is commonly even more tenuous in the domain of adaptation, where the link between company actions and organisational or even societal adaptation is often more indirect and ambiguous (a feature characterising the adaptation debate more broadly, as well; Ziervogel and Taylor, 2008). In such circumstances, our test for inclusion in most of the empirical chapters of this book, is whether the company or companies concerned has or have an explicit intention to contribute to either mitigation or adaptation. The latter includes organisational changes targeted at improved organisational resilience (Berkhout et al., 2006), as well as externally orientated efforts to enhance communities' and ecosystems' resilience to climate changes (Folke et al., 2002). Sometimes, however, we include phenomena where even tentative intentions to respond to climate change are difficult to show – this is because the link to impacts related to climate change is conceptually clear and/or the phenomena in question hold important lessons for mitigation or adaptation. (To illustrate, none of the business owners interviewed by Shale in Chapter 9 had climate change in mind when discussing risks associated with flooding, but their responses to this risk are clearly pertinent in the context of expected increases in floods in Cape Town.)

Climate change as a 'wicked problem' and areas of limited statehood

Climate change has been characterised as a 'wicked problem', which 'defies resolution because of the enormous interdependencies, uncertainties, circularities, and conflicting stakeholders implicated by any effort to develop a solution' (Lazarus, 2009: 1160; see also Australian Public Service Commission, 2007). Climate change related phenomena and processes epitomise the complex, non-linear interactions between social and natural systems (Liu et al., 2007; see also Anderies et al., 2004), which give rise to high degrees of unpredictability and path dependence. Cognitive challenges in understanding climate change related processes and risks are further exacerbated by politically charged disputes surrounding climate science. As a result, even though there is a relatively significant degree of agreement among scientists about key aspects of climate science, there remain significant uncertainties and variation in public understanding of the science and policy options related to climate change.

These cognitive challenges are met by economic and political challenges related to devising rules and establishing prices for costs and benefits that are often very distant in time and space. Climate change is thus a classic 'collective action' problem, because individuals' incentives to constrain their actions are diffuse and uncertain in comparison to the benefits of inaction, resulting in significant motives to 'free-ride' – that is, to rely on others (it is hoped) to make the necessary changes, while resisting changes oneself. This is further exacerbated by the high degree of consumer preference for products and services implicated in causing climate change – fossil fuels in particular – and the broader socio-technical regimes that rely on these fuels and associated modes of production and transport. As a result, even if prices rise, consumption will probably remain high. Recognising the interrelated and co-dynamic nature of technological, institutional and organisational systems that underpin the climate change problem, van den Berg and colleagues (2011: 13) emphasise the need for a 'new generation of governance approaches' with a 'long-term orientation' and corresponding policy design that is 'flexible, adaptive and reflexive', with an emphasis on 'deliberation, probing (exploration scenarios), experimentation and learning'. In a related vein, Brunner and Lynch (2010) highlight constraints faced by states to coordinate effective responses to climate change and emphasise a shift towards 'bottom-up' responses by community groups, social movements and even businesses acting voluntarily in enlightened self-interest or moral compunction.

Yet the governance literature suggests that such voluntary action by business is often incentivised by the 'shadow of hierarchy'. In order to avoid state regulation, firms may choose voluntarily to commit themselves to reaching a regulatory outcome closer to their preferences. Moreover, the possibility of state intervention reduces the incentive to renege on a voluntary commitment (cf. Mayntz and Scharpf, 1995b; Scharpf, 1997; Héritier and Lehmkuhl, 2008). So even if climate change presents states with fundamental challenges to regulate effectively, the credible threat by states to embark on such regulatory action is likely to be an important driver for firms to adopt self- or co-regulatory initiatives. But what about those geographic areas in which states have only limited capacity to create this credible threat because 'political institutions are too weak to hierarchically adopt and enforce collectively binding rules' (Börzel and Risse, 2010: 13)?

There is empirical evidence that suggests that companies do contribute to climate change governance in such geographic areas of limited statehood, despite the absence of a credible shadow of hierarchy. Companies do not only adopt global standards to govern their worldwide business activities (Epstein and Roy, 2007; Prakash and Potoski, 2006, Prakash and Potoski, 2007). They also voluntarily implement environmental protection standards or agree to use sustainable energy (Flohr et al., 2010). In some instances, they even regulate their supply chains (Héritier et al., 2009) and seek to foster state regulation, by pressing for stricter legislation and helping to strengthen the enforcement capacity of state actors (Vogel and Kagan, 2004; Mol, 2001; Flanagan, 2006; Börzel and Thauer, 2013). Despite relatively abundant empirical research on business responses to climate change in developed countries (Begg et al., 2005; Kolk and Pinkse, 2009; Levy and Kolk, 2002; Sullivan, 2009), there has been little research on this more specific theme in developing countries, particularly in Africa. Existing research often targets practitioners and policy makers as the primary audience and hardly engages with related scholarly debates. One of the objectives of this book is to help to fill this research gap.

Much of our empirical analysis focuses on South Africa, though we use some evidence from Kenya and Germany for comparison. As in other sub-Saharan countries, climate change is becoming a growing development and human rights concern in South Africa, threatening business as usual, especially for the poor. Changing weather patterns and resulting ecological and social effects, including reduced crop productivity and water availability (Bates et al., 2008; Kranz, 2010), and associated impacts on health, food security, and the economy, are expected to be especially adverse in this region. For most sub-Saharan African countries, these climate change impacts are cruelly disproportionate to their histori-cal and current GHG emissions. South Africa is something of an anomaly, due to its relatively high per capita GHG emissions, despite widespread poverty. Across the region, high levels of poverty and inequality, coupled with low levels of state capacity – shown by expansive informal settle-ments and weak education and public health systems – all contribute to disturbing constraints on these countries' ability to effectively respond and adapt to climate change. At the same time, pressing socio-economic development needs are generally a priority, and these sometimes align and sometimes conflict with climate mitigation and adaptation efforts (Ziervogel and Taylor, 2008).

South Africa provides a fruitful setting for the study of business and climate change governance in areas of limited statehood, with significant potential to extrapolate to other emerging economies. South African companies face significant pressures related to climate change, including the need to reduce GHG emissions and to adapt to projected decreases in water availability. At the same time, they are operating in a governance context characterised by highly variable state capacity, with ambitious GHG reduction policies at the national level, on the one hand, and often far-reaching constraints to local municipalities' provision of even basic infrastructure on the other. South Africa also allows for comparison of a wide variety of companies, ranging from large, multinational compa-nies to smaller, domestic firms and even survivalist micro-enterprises. The empirical chapters in this book make good use of these variations across problem characteristics, institutional conditions and organisational aspects. Where the comparative scope requires further expansion, with emphasis on the institutional dimension, we include empirical material from Kenya (in Kapfudzaruwa's survey of corporate reports (Chapter 2)) and Germany (in Fay's analysis of cost of capital in wind energy projects (Chapter 4) and in Kranz's discussion of water management (Chapter 6)).

Characterising how companies contribute to climate change governance

Much of the scholarly work has focused, often implicitly, on the role of business organisations in mitigation, rather than adaptation. This is also because there has been comparatively less emphasis on adaptation among businesses, which in turn may be due to the lack of a 'com-mon definition of what adaptation means for business' (Pinkse and Kolk, 2012: 182; see also Berkhout et al., 2006). This might be due to the abovementioned focus of current research on business and climate change in developed economies. The contributions to this volume clearly demonstrate that businesses are contributing in a variety of ways to adaptation in developing or emerging economies.

Another objective of this book, therefore, is to give equal and simultaneous attention to business responses to both mitigation and adaptation, recognising that while these often represent distinct policy domains, they also entail important areas of overlap and complementarities, especially in developing or emerging economy countries. In the following paragraphs, we will thus briefly profile existing typologies of companies' responses in either mitigation or adaptation, in order to then develop what we hope is a parsimonious framework to characterise both.

Business contributions to climate change mitigation

Kolk and Pinkse's (Kolk and Pinkse, 2005) typology of companies' strategic responses to climate change focuses on mitigation. They argue that companies' responses focus on two overarching strategic intentions: *innovation* and *compensation*. Innovation involves the development of environmental technologies or services to reduce emissions while compensation leaves a company's own technological competencies unaltered as the company uses emission reduction technologies developed by other companies. When these two overarching strategic aims are combined with different levels of organisational activities and interactions, a matrix is developed to outline the strategic options used in response to climate change (see Figure 1.2). Six strategic options emerge that could be part of a comprehensive strategy for climate change in which companies combine several options.

With regard to *innovation* options, process improvement is the main way in which companies' make efforts to reduce not only emissions but also input costs of energy and materials (Porter and van der Linde, 1995). Innovation processes can also focus on new product development, commonly involving other role-players in the supply chain, and driven in large part by increasingly prominent consumer concerns about climate change and their willingness to consider full product life cycles (Nidumolu et al., 2009). Finally, companies may collaborate beyond traditional supply chains for the development of new product and market combinations: 'The cooperation between oil and automobile companies on fuel cells is a case in point' (Kolk and Pinkse, 2005: 9).

Compensation inspired responses include in the first instance GHG accounting efforts. Measuring and publicly reporting the organisational GHG emissions is a widely expected first step, followed by setting targets for reducing GHG emission. Another internal company compensation

		Main aim	
		Innovation	Compensation
Organisational activities & interactions	Internal (company)	Process improvement	GHG accounting and internal transfers
	Vertical (supply chain)	Product development	Supply-chain measures
	Horizontal (beyond the supply chain)	New product and market combinations	Acquisition of emission credits

Figure 1.2 Typology of strategic options for corporate responses to climate change (with a focus on mitigation), according to Kolk and Pinkse (2005)

strategy is to transfer GHG intensive activities to parts of the company that operate in less stringent jurisdictions (Kolk and Pinkse, 2009; Hoffman, 2007). Such transfers can also be effected within the supply chain, for instance by outsourcing particularly GHG intensive activities, such as transportation, or by choosing 'cleaner' electricity or raw products. Finally, compensatory responses may entail purchasing carbon credits on either the compliance or voluntary carbon markets.

Kolk and Pinkse's framework is a useful point of departure, but of course it is not comprehensive. For instance, it does not consider companies' political activism in order to influence public policy or market-based incentives, and neither does it capture business action within associations, such as the South African Energy Efficiency Initiative discussed by Klapper and Kahn in Chapter 5 in this book. Furthermore, it does not cater for business adaptation efforts, to which we now turn in order to propose an integrative framework.

Business contributions to climate change adaptation

Moser and Ekstrom suggest that adaptation strategies and actions 'can range from short-term coping to longer-term, deeper transformations, aim to meet more than climate change goals alone, and may or may not succeed in moderating harm or exploiting beneficial opportunities' (Moser and Ekstrom, 2010: 1). These options are schematically illustrated in Figure 1.3, with the amount of resources and effort required on the y-axis, and the predominant time-frame on the x-axis. An organisation may be active on different points on the spectrum suggested in Figure 1.3, just as it may mix different options in the mitigation typology discussed above. However, the more it seeks to affect system transformation, the more resources, time and effort it is likely to require. This is also because the external barriers to system transformation are likely to be higher. Furthermore, each type of adaptation strategy requires a different set of adaptive capacities, with system transformation likely to demand greater levels of learning and flexibility.

Coping strategies have been defined as 'temporary responses to a familiar disturbance or transient threat' (Christopolos et al., 2009: 15). They are often ad hoc and rely on existing assets and capabilities. This type of adaptation is mainly pursued by organisations that take a 'wait and see' approach, which is described by Berkhout et al. as 'a strategy of deferral, based on skepticism or uncertainty about the possible impacts of climate change' (Berkhout et al., 2006: 151). Substantial adjustments, on the other hand, imply 'a process of planning leading to a lasting change in the risk environment' (ibid.). They involve longer-term adjustments including changes in behaviour, practices and values. System transformation refers to a fundamental restructuring of system components and relationships (Walker et al., 2004; Nelson, 2010). System thresholds separating fundamental system states can be crossed unintentionally through collapse, or intentionally through deliberate efforts (Nelson et al., 2007), and in the context of adaptation the focus is of course on the latter. Our focus, therefore, is on actions taken by business actors

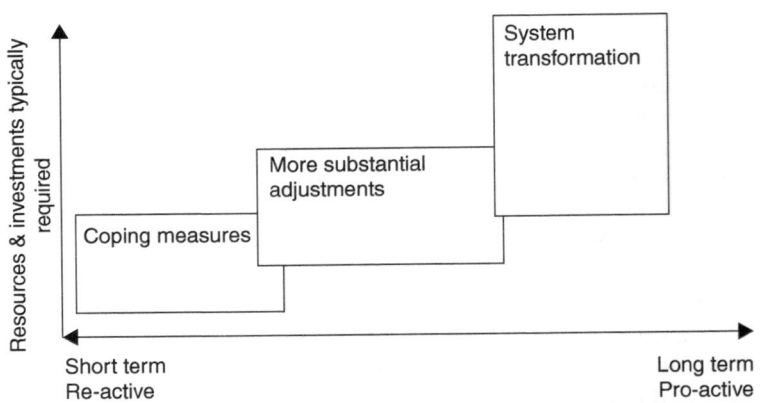

Figure 1.3 Adaptation options ranging from coping to transformation, based on Moser and Ekstrom (2010)

to restructure fundamentally the interactions between system constituents, including other social actors, resources and relationships.

An integrated model of firm responses to climate change

Each of the typologies introduced above emphasises different dimensions according to which firm activities are categorised. Kolk and Pinkse (2005) emphasise the strategic intention of mitigation actions and the level at which they are implemented, while Moser and Ekstrom (2010) focus on resources and effort exerted, on the one hand, and the time-scale underlying the rationale of these investments, on the other. We suggest that it is feasible and helpful to develop an integrated typology drawing on dimensions developed in the above two frameworks, and which can cater for both mitigation and adaptation efforts. The first dimension, borrowed from Moser and Ekstrom, recognises that short-term time-frames used to rationalise adaptation efforts are associated with responses that generally require fewer resources and effort. We suggest that this amalgam between time-frame and effort applies also to mitigation – shorter-term analysis of mitigation pressure will result in 'low-hanging fruit' approaches to GHG emission reductions, with an emphasis on the use of existing capabilities. In effect, the combination of level of effort (in other words, the relative size of investments) and the expected time-frame of returns gives rise to what Thauer calls the 'asset specificity' of investments (Chapter 3).

The second dimension pertains to whether the capabilities and resources developed and applied in the firm's response are predominantly within or outside the firm's organisational boundaries. This is of course closely related to the levels of mitigation action used in Kolk and Pinkse's framework. It applies to adaptation, too, because firms can bear and manage climate-related risks within their own boundaries, or share and shift risks with or to other role-players (Berkhout et al., 2006). Combining these dimensions results in a framework depicted in Figure 1.4, with seven categories of business responses located relative to the axes of decision-making timeframes and the focus of efforts within or outside firm boundaries.

The basic framework provided in Figure 1.4 can also be used to map the types of business initiatives that are focused on in the empirical analyses in this book, as indicated in Figure 1.5:

- Kapfudzaruwa (Chapter 2) uses Kolk and Pinkse's categories to identify strategic configurations among listed South African and Kenyan companies' responses to climate change mitigation. That is, he is looking at efforts that cover a broad spectrum across the decision-making time-frame and locus of effort axes.

- Thauer (Chapter 3) focuses on the motives for process improvements in the operations of multinational car makers in South Africa.
- Fay (Chapter 4) discusses the financial considerations of developers of renewable energy projects – that is, new product providers – in a comparative study of hypothetical projects in South Africa and Germany in order to assess the financial implications of diverse levels of statehood.
- Kaan and Klapper (Chapter 5) consider the motives for firms' participation in a collective agreement, the South African Energy Efficiency Initiative. That is, their focus is not on companies' efforts at improving energy efficiency within their organisations per se (which would primarily represent process improvements akin to those considered by Thauer in Chapter 3), but rather their participation in a collective initiative to ensure broader commitment and action.
- Herbstein and colleagues (Chapter 8), Kranz (Chapter 6), Methner (Chapter 7) and Shale (Chapter 9) each discuss various means and motives for climate change adaptation among diverse companies – ranging from large retailers (as in Methner's case study) to informal, survivalist enterprises in Cape Town's townships (Shale). Kranz and Methner explicitly discuss the spectrum of adaptation responses suggested by Moser and Ekstrom. Herbstein and colleagues consider the challenges of moving toward more systemic, transformational efforts in insurance companies. Shale, meanwhile, illustrates why informal insurance is not adopted by survivalist businesses in low-income urban communities in Cape Town, South Africa, with business owners and residents in these areas rather relying on associations based on religious beliefs and norms of reciprocity. This suggests an approach that is immune not only to the marketing efforts of large, formal sector insurance companies, but is markedly different also to expectations among most scholars investigating organisational adaptation to climate change.

Framing the drivers and conditions of companies' contributions to climate change governance

Problem salience and characteristics

It is apparent that companies in different sectors will face different pressures and opportunities for both climate change mitigation and adaptation. Focusing on mitigation, Kolk and Pinkse identify three categories of firms in terms of how they are exposed to climate change (Kolk and Pinkse, 2008, Kolk and Pinkse, 2012). Their category descriptions are

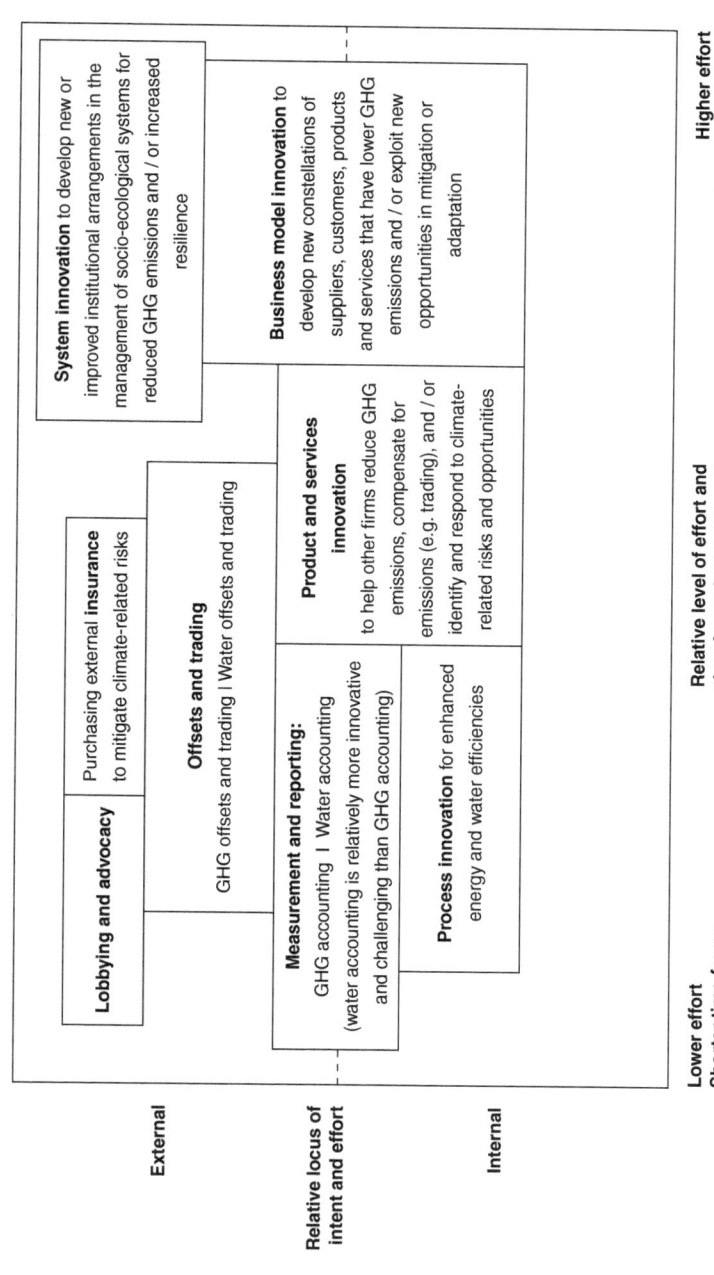

Figure 1.4 An integrated framework of firm responses to climate change mitigation and adaptation, with a focus on energy- and water-related efforts

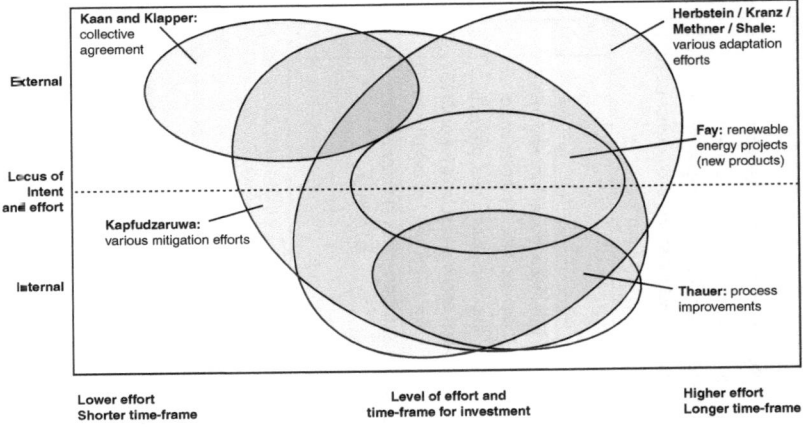

Figure 1.5 Using the integrated framework of firm responses to climate change to map empirical analyses in this book

adapted in the 'mitigation' column in Table 1.1. Firms in 'high-salience sectors', such as oil and gas, automobiles and utilities, have their core activities at stake, 'with their fossil-fuel based business models being threatened' (Kolk and Pinkse, 2012: 3), and can thus reap competitive advantages through early changes in business models. Firms specialising in climate-relevant goods and services, such as consultants, banks and renewable energy companies, can benefit by helping other firms respond effectively and by operating in carbon markets (of which the European Emissions Trading Scheme is the most prominent). The remainder of firms are characterised by relatively lower emissions and may engage in climate mitigation activities in order to enhance their reputation.

Similar categories are likely to be relevant in the adaptation domain, as noted in the 'adaptation' column in Table 1.1. Firms active in agriculture and food production or retail, for instance, are especially exposed to changes in water availability or extreme events, such as droughts and floods, and the same applies to the insurance sector. These thus represent high-salience sectors with regard to climate change adaptation. Firms that provide, for instance, water purification products or drought-resistant seeds, may be characterised as firms specialising in adaptation-relevant goods and services. Finally, some firms may not consider themselves to be exposed to climate change related risks, but this is probably less clear than in the case of emissions-intensive companies and mitigation concerns, because the interrelated nature of

Table 1.1 Relevance of climate change (mitigation) for different categories of firms (adapted from Kolk and Pinkse, 2012)

Category of Firms	Impacts of climate change issues	
	Mitigation	**Adaptation**
Firms in high-salience sectors	• Strongly affected in view of energy intensity and dependence and/or emissions intensity • Early change in business models might be source of competitive advantage in terms of changing state policies, markets, and consumer preferences	• Strongly affected in view of climate-related risks, such as water scarcity or extreme weather events • Early change in business models might be source of competitive advantage in terms of expected climatic changes and systemic effects
Firms specialised in climate-relevant goods and services	• Can profit by helping firms mitigate GHG emissions or to anticipate, influence or respond to climate policy	• Can profit by helping firms or other organisations adapt to expected climatic changes and systemic effects
Remaining firms with low-emission activities or limited exposure to climate change risks	• No main (perceived) source of profitability/growth; may gain legitimacy from acting visibly • Address issues via external markets; possibility for internalisation arbitrage	• No main perceived source of profitability/growth; may gain legitimacy from acting visibly

climate and socio-economic systems will give rise to impacts on most if not all sectors.

A company's sector is thus likely to influence the degree and manner in which it engages in climate change governance. However, there is a range of business responses associated with climate change governance (see Figure 1.4). Across these various options, it is likely that we can expect different motives and drivers to become more prevalent. More specifically, the relative importance of state regulation or the threat thereof is likely to be more pertinent in some areas of Figure 1.4 than in others. At the same time, the absence of credible state action may be seen to contribute to a governance gap that could impede business success, thus motivating business contributions to governance in some response categories. These different drivers are discussed in more detail below, but the point to be made is that the characteristics of the problem, specifically its task complexity, are likely to influence the relative importance of such drivers.

Institutional and firm-level drivers

Management scholars have emphasised the efficiency gains to be achieved through process improvement and the potential for increased market share through the development of environmentally friendly products (Ammenberg and Hjelm, 2003; Anton et al., 2004; Parker, 2002; Porter and Kramer, 2002, Porter and van der Linde, 1995; Barney, 1997). Innovation strategies motivated by process efficiencies, new product markets, or even new business models are increasingly characterised as the holy grail of sustainability strategy, with the potential to create 'shared value' for society and the innovating business (Porter and Kramer, 2011).

Yet such cutting edge strategies are perhaps beyond most companies, and they are also likely to be especially challenging in particular sectors (such as mining, for example). Furthermore, as mentioned above, current literature on business responses to climate change has emphasised the significance of regulation, or the threat of regulation, as a driving force. This suggests a prominent role for institutional theory, which emphasises the pressure on companies to respond to external pressures in an effort to gain or maintain legitimacy in a company's operating environment (Deegan, 2007; Dowling and Pfeffer, 1975; Meyer and Rowan, 1977; Zucker, 1977). In practice, of course, firm strategies are motivated by both competitiveness and legitimacy, and these drivers are often interrelated or enmeshed. Rather than focus on either of these as distinct categories of drivers, we will emphasise two levels of analysis – the institutional level and the organisational level – in order to help explain why some companies respond to climate change in different ways than others.

At the institutional level, the importance of state regulation, or the absence thereof, has already been emphasised. What remains to be said is that regulative institutions have diverse ways of influencing companies' actions beyond straightforward 'command-and-control' compulsion. One aspect of this relates to the role of home government regulation: while high-regulating countries are reluctant to regulate their companies outside their territory (the Alien Tort Claims Act of the US is a prominent exception (Deitelhoff et al., 2010)), we know that Multinational Corporations (MNC) tend to transport their regulatory standards abroad as these are interpreted as 'quality signals' (Potoski and Prakash, 2009) by business partners and customers (Murphy, 2000; Skjaerseth and Skodvin, 2003; Hall and Soskice, 2001; Xing and Kolstad, 2002; Kolk et al., 1999). Furthermore, a firm with high standards faced with a competitor with lower standards has an incentive to press the government to issue stricter regulation (Börzel et al., 2011).

Regulatory institutions are generally characterised by change and uncertainty in emerging institutional fields, and this is also the case in the climate change domain, even more so in areas of limited statehood. Such circumstances often give rise to 'mimetic' pressures at the institutional level, as companies seek to emulate companies considered innovative and successful (DiMaggio and Powell, 1983; Bansal, 2005; Potoski and Prakash, 2005; Potoski and Prakash, 2006). Imitation and peer pressure may thus become important elements in understanding firms' responses to climate change. Furthermore, business associations, collective agreements and informal networks often act as transmitters of peer pressure (Kollman and Prakash, 2001), further contributing to an increasingly homogeneous set of organisational responses in a process called mimetic isomorphism (DiMaggio and Powell, 1991). However, empirical work suggests that isomorphism is not the dominant trend in organisations' responses to climate change (Kolk and Pinkse, 2005), so a better understanding of how different institutional settings exert differential pressures on different kinds of organisations is required.

Secondly, governance institutions may be playing an important role not only by being present, but also by being absent, and thus giving rise to a 'shadow of anarchy' (Mayntz and Scharpf, 1995a). So, while companies operating in areas of limited statehood face a relatively less credible shadow of hierarchy cast by the host state, it may be precisely the absence of the threat of strict(er) regulation or public service provision that creates an incentive for companies to foster regulation or provide public services. If the state is not capable of adopting and enforcing collectively binding decisions, companies are not confronted with a situation in which they have to weigh the costs of cooperation and voluntary commitment against the possibility of a suboptimal hierarchically imposed policy. Rather, they face the danger of not having a common good at all. If the pursuit of their individual profit depends on the provision of certain common goods and collectively binding rules to produce them, and the state is not capable or unwilling to provide them, the 'shadow of anarchy' provides companies with a major incentive to step in and fill the governance gap (Ruggie, 2004b; Börzel, 2010).

Finally, a range of normative institutions have been emerging in the organisational field of business and climate change, including also related fields such as corporate responsibility more generally. For instance, the United Nations Global Compact, the Global Reporting Initiative, or ISO 14001 commit companies to social and environmental standards or a voluntary basis, even if some of these 'soft law' instruments have become de facto requirements through some states' laws or companies' supply chain buying criteria. These normative pressures are likely to play an important role in both mitigation and adaptation activities, although in diverse ways. The mitigation agenda has received very significant civil society and consumer attention. This is likely to have influenced companies, especially those with prominent brand names and large GHG emissions, such as energy or petrochemical companies. To some extent, such mitigation-orientated norm activism may be present in local communities surrounding business operations, but it is likely to be prominent only if other, more direct environmental risks are felt by such communities, as is the case surrounding large industrial operations of companies such as Shell or Acelor Mittal in South Africa.

The same is likely to be true for corporate adaptation actions. Pressures by local communities or civil society organisations for companies to help provide water supply infrastructure or reduce water pollution are likely to be motivated primarily by more short-term, direct interests related to water consumption, rather than longer-term links between climate change and water scarcity (even if the latter is included prominently in such groups' communication tactics).

From the above, it is clear that firm-level characteristics play an important role in influencing how companies will respond to institutional pressures or market opportunities related to climate change. For instance, the vulnerability of companies to public pressure is likely to be stronger if a company has a brand name to protect, if it targets a high-end market (Haufler, 2001; Mol, 2001: 97–100; Blanton and Blanton, 2007), if it has an international (export) orientation (Bansal and Roth, 2000), or if its product is highly visible to end-consumers (Deitelhoff and Wolf, 2010). Many of these findings rely on the importance of company reputation as a key corporate asset (Spar and LaMure, 2003).

Further firm-level drivers relate to organisational culture and leadership, and numerous scholars and commentators highligh their role in motivating and facilitating organisational changes in response to climate change and other sustainability concerns. For instance, Kolk and Levy describe the impact of a new generation of leaders in oil multinationals (Kolk and Levy, 2001). Another particularly important organisational capability that has been emphasised, also with specific reference to climate change, pertains to knowledge and organisational learning (Berkhout, 2011). This relates to companies' ability to sense the possible implications of climate change for their business models, and some have highlighted the role of companies' interactions with other stakeholders in this process of learning

(Aragon-Correa and Sharma, 2003; Hart and Sharma, 2004). More generally, the dynamic capabilities perspective in the management literature and its focus on companies' ability to learn – 'learning to learn' – has been emphasised also in the sustainability and specifically the climate change field (Berkhout, 2011; Hart and Dowell, 2011).

An integrated, preliminary model of drivers and overview of chapters' key contributions

In the previous two sub-sections, the literature survey has emphasised three primary categories of drivers and conditions that are likely to help explain different levels and manifestations of business contributions to climate change governance: problem characteristics, organisational factors and institutional drivers. While much of our discussion has emphasised the interrelationships between these motives and conditions, it is nevertheless useful to consider these categories separately in our preliminary model in order to better investigate their role and interrelationship in the empirical analyses discussed in this book. These categories and some of the main conceptual elements are shown in Figure 1.6. Their relative importance and interactions will feature prominently in each of the empirical analyses that follow, though in some instances (as in Thauer's chapter) they are augmented by other factors):

- · Kapfudzaruwa (Chapter 2) finds that within countries, divergence in companies' efforts is to some extent explained by their sectors (i.e. problem salience and characteristics) and a second source of divergence relates to companies' capabilities and leadership (i.e. organisational drivers). The significant variation between South Africa and Kenya, in companies' responses, points to an important role for institutional factors, specifically the maturity of the climate change organisational field and the relatively stronger state in South Africa.
- Thauer (Chapter 3) demonstrates that expected institutional or organisational drivers cannot account for divergence in multinational car manufacturer's process improvement standards. Rather, he argues that this can be explained by variations in the extent to which companies have made long-term investment commitments in their South African operations (i.e. variations in asset specificity of investments).
- Fay (Chapter 4) discusses the various factors that contribute to the cost of capital in renewable energy projects and finds that issues related to limited statehood contribute to the significantly higher cost of capital for projects in South Africa, giving rise to important challenges for policy makers and project developers.

- Klapper and Kaan (Chapter 5) assess the reasons why companies participate in the South African Energy Efficiency Accord. Their findings suggest the unexpected possibility that the 'shadow of hierarchy' and the 'shadow of anarchy' can play a role concurrently, because of the varying task complexity of different aspects of the problem. On the one hand, businesses participate because of their concerns about energy security, i.e. a lack of faith in the state's ability to ensure reliable energy supply. On the other, they also engage in the initiative in order to manage risks associated with possible state regulation to enforce cuts in energy consumption. That is, the shadow of anarchy is present in the complex area of national energy security, while the shadow of hierarchy looms in the potential for compulsory energy cuts or emission taxes.
- Kranz (Chapter 6) considers business contributions to climate change adaptation and adaptive water management in South Africa and Kenya. She finds that such efforts are more prominent in South Africa, which is largely due to the greater problem pressure in that country, as well as business decision-makers' perceptions of limited state capacity to deal with complex water management problems. Conversely, German business decision-makers are holding off from proactive efforts because they expect hierarchical state action, which may well be an impediment to theoretically motivated and policy-related ambitions to enhance more adaptive and participatory forms of governance in that context.
- Methner (Chapter 7) continues this investigation into business contributions to adaptation with an in-depth case study of a South African retailer known for its strategic and proactive sustainability commitments. She finds that this company's leadership position can be explained in large part by the confluence of organisational and institutional drivers – on the one hand, the company's strategy is premised on maintaining a distinctive brand focused on higher income consumers and long-term relationships with suppliers; on the other, it recognises the need to respond to resource constraints, such as declining soil fertility and water availability, in the context of limited state action.
- Herbstein and colleagues (Chapter 8) note that there are important incentives for insurance companies to develop more systemic, transformational responses to climate risk, going beyond incremental changes in their business model to address the proximate drivers of risk in collaboration with other role-players. These incentives are related to the sector-specific salience of the problem pressures associated with growing risk of climate-related extreme weather events, as well as an apparent inability of the state to address these.

At the same time, the authors identify a variety of organisational and institutional obstacles to making such changes.

- Shale (Chapter 9) focuses on a vital set of conditions in the context of developing and emerging economy countries: small, informal and often survivalist businesses operating in markets, in which larger, formal businesses struggle to obtain a foothold despite ongoing efforts at 'base of the pyramid' marketing. She finds that such business owners largely avoid formal insurance in their response to climate-related hazards – specifically the regular floods affecting such areas in Cape Town – but rather participate in voluntary associations and particularly burial societies, which provide vital social networks and respond to religious and cultural values of their participants.

Conclusion

In this introductory chapter, we have outlined a conceptual framework for investigating how and why business organisations contribute to

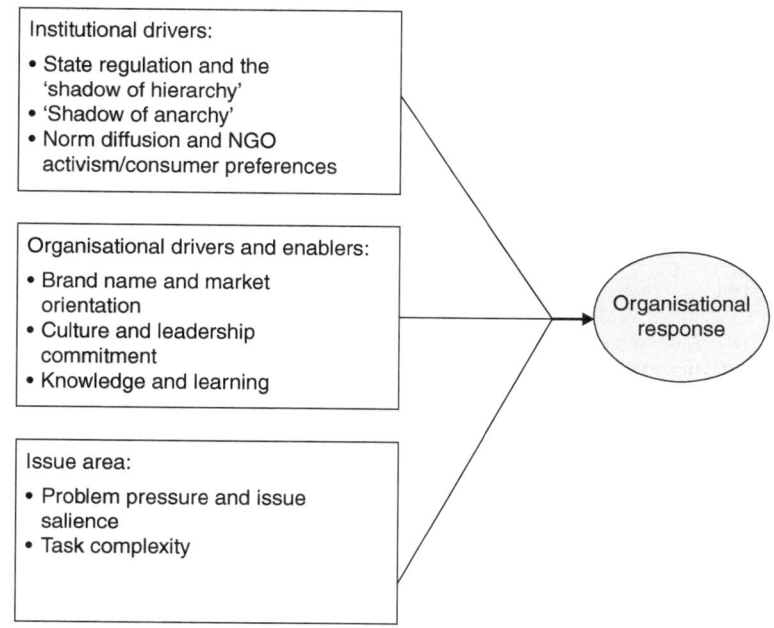

Figure 1.6 Preliminary model of expected drivers and conditions for business contributions to climate change governance

climate change governance. The organisational field emerging around the climate change issue domain provides various pressures and incentives for business responses, and at the same time it is open to influence from firms acting independently or collectively. We characterised firms' responses to climate change by developing a framework that emphasises the decision-making timeframe and relative size of investments and the location of effort. This framework indicates the diversity of business contributions to climate change governance. Some of the chapters in this book cover a broad array of initiatives across these dimensions, while others focus on more particular areas and aspects.

We then discussed the key drivers and conditions likely to influence firms' contributions to climate change governance, resulting in a preliminary model emphasising problem characteristics, organisational factors and institutional drivers. The empirical chapters highlight the relative importance and interrelationship of these particular drivers and conditions, which are further discussed in the concluding chapter, together with organisational strategy and policy-related implications. One of the overarching, unexpected conclusions is that the 'shadow of hierarchy' and 'shadow of anarchy' can be present concurrently to motivate business responses, because of the varying degrees of task complexity of options or responsibilities faced by the state. This suggests a more nuanced treatment of the interplay between institutional and organisational drivers for proactive sustainability strategies among companies is necessary in the literature, and it also broadens our view of options for motivating business contributions to climate change governance.

Notes

1. http://www.guardian.co.uk/environment/2007/nov/29/climatechange. carbonemissions, last accessed June 15, 2012.
2. http://www.cbi.org.uk/media/1044924/green_business_dinner_-_christiana_ figueres.pdf, last accessed June 15, 2012.
3. These two initiatives focus attention on the need for companies to report publicly on their policies and practices on sustainability issues such as climate change, with the latter focusing in particular on greenhouse gas emissions.

References

Ammenberg, J. and Hjelm, O. (2003) 'Tracing Business and Environmental Effects of Environmental Management Systems – A Study of Networking Small and Medium-Sized Enterprises Using a Joint Environmental Management System', *Business Strategy and the Environment* 12, 3, 163–174.

Anderies, M., Janssen, M. and Ostrom, E. (2004) 'A framework to analyze the robustness of social-ecological systems from an institutional perspective' 9, 1, http://www.ecologyandsociety.org/vol9/iss1/art18/.

Anton, W.R., Deltas, G. and Khanna, M. (2004) 'Incentives for Environmental Self-Regulation and Implications for Environmental Performance', *Journal of Environmental Economics and Management* 48, 1, 632–654.

Aragon-Correa, A.J. and Sharma, S. (2003) 'A Contingent Natural-Resource Based View of Proactive Environmental Strategy', *Academy of Management Review* 28, 1, 71–88.

Australian Public Service Commission (2007) *Tackling Wicked Problems: A Public Policy Perspective* (Canberra: Australian Public Service Commission).

Bansal, P. and Roth, K. (2000) 'Why Companies Go Green: A Model of Ecological Responsiveness', *The Academy of Management Journal* 43, 4, 717–736.

Bansal, P. (2005) 'Evolving Sustainably: A Longitudinal Study of Corporate Sustainable Development', *Strategic Management Journal* 26, 3, 197–218.

Barney, J.B. (1997) *Gaining and Sustaining Competitive Advantage* (Reading, MA: Addison-Wesley).

Bates, B., Kundzewicz, Z., Wu, S. and Palutikof J. (eds) (2008) *Climate Change and Water: Technical Paper of the Intergovernmental Panel on Climate Change* (Geneva: IPCC Secretariat).

Begg, K.G., van der Woerd, F. and Levy, D.L. (2005) *The Business of Climate Change* (Sheffield, UK: Greenleaf).

Berkhout, F. (2011) 'Adaptation to Climate Change by Organizations', *WIREs Clim Change* 3, 91–106.

Berkhout, F., Hertin, J. and Gann, D. (2006) 'Learning to Adapt: Organizational Adaptation to Climate Change Impacts', *Climatic Change* 78, 135–156.

Blanton, R.G. and Blanton, S.L. (2007) 'Human Rights and Trade: Beyond the "Spotlight"', *International Interactions* 33, 2, 97–117.

Börzel, T.A. (2010) 'Governance with(out) Government – False Promises or Flawed Premises', *SFB Working Papers; Sonderforschungsbereich 700, Freie Universität Berlin*, 23.

Börzel, T.A. and Risse, T. (2010) 'Governance without a State – Can It Work?', *Regulation and Governance* 4, 2, 1–22.

Börzel, T.A., Héritier, A., Kranz, N. and Thauer, C. (2011) 'Racing to the Top? Regulatory Competition among Firms in Areas of Limited Statehood', in T Risse (ed.) *Governance without a State? Policies and Politics in Areas of Limited Statehood* (New York: Columbia University Press).

Börzel, T.A. and Thauer, C. (eds) (2013) *Racing to the Top? Business and Governance in South Africa* (Houndmills: Palgrave Macmillan).

Brühl, T., Debiel, T., Hamm, B., Hummel, H. and Martens, J. (eds) (2001) *Die Privatisierung der Weltpolitik. Entstaatlichung und Kommerzialisierung im Globalisierungsprozess* (Bonn: Dietz).

Brunner, R.D. and Lynch, A.H. (2010) *Adaptive Governance and Climate Change* (Boston, MA: American Meteorological Society).

Burris, S., Hancock, T., Lin, V. and Herzog, A. (2008) 'Emerging Principles of Healthy Urban Governance', *Thematic Paper 5 of the Knowledge Network on Urban Settings, WHO Centre for Health Development*.

Chan, A. and Ross, R.J. (2003) 'Racing to the Bottom. Industrial Trade without a Social Clause', *Third World Quarterly* 24, 6, 1011–1028.

Christopolos, I., Klein, R.J.T., Anderson, S., Le Goulven, K., Arnold, M., Galaz, V. and Hedger, V. (2009) *The Human Dimension of Climate Adaptation: The Importance of Local and Institutional Issues* (Stockholm: The Commission on Climate Change and Development).

Deegan, C. (2007) 'Organizational Legitimacy as a Motive for Sustainability Reporting', in J Unerman, J Bebbington and B O'Dwyer (eds) *Sustainability Accounting and Accountability* (London: Routledge).

Deitelhoff, N., Feil, M., Fischer, S., Haidvogel, A., Wolf, K.D. and Zimmermann, M. (2010) 'Business in Zones of Conflict and Global Security Governance: What Has Been Learnt and Where to from here?' in N Deitelhoff and KD Wolf (eds) *Corporate Security Responsibility? Corporate Governance Contributions to Peace and Security in Zones of Conflict* (London: Palgrave).

Deitelhoff, N. and Wolf, K.D. (2010) (eds) *Corporate Security Responsibility? Corporate Governance Contributions to Peace and Security in Zones of Conflict* (Houndmills: Palgrave).

DiMaggio, P.J. (1991) 'Constructing an Organizational Field as a Professional Project', in WW Powell and PJ DiMaggio (eds) *The New Institutionalism in Organizational Analysis* (Chicago: Chicago University Press).

DiMaggio, P.J. and Powell, W.W. (1983) 'The Iron Cage Revisited: Institutional Isomorphism and Collective Rationality in Organizational Fields', *American Sociological Review* 48, 2, 147–160.

DiMaggio, P.J. and Powell, W.W. (1991) 'Introduction', in WW Powell and PL DiMaggio (eds) *The New Institutionalism in Organizational Analysis* (Chicago: Chicago University Press).

Dowling, J. and Pfeffer, J. (1975) 'Organisational Legitimacy: Social Values and Organisational Behaviour', *Pacific Sociological Review* 18, 1, 122–136.

Epstein, M.J. and Roy, M.-J. (2007) 'Implementing a Corporate Environmental Strategy. Establishing Coordination and Control within Multinational Companies', *Business Strategy and the Environment* 16, 6, 389–403.

Flanagan, R.J. (2006) *Globalization and Labour Conditions* (Oxford: Oxford University Press).

Flohr, A., Rieth, L., Schwindenhammer, S. and Wolf, K.D. (2010) *The Role of Business in Global Governance. Corporations as Norm-Entrepreneurs* (Basingstoke: Palgrave).

Folke, C., Carbenter, S., Elmqvist, T., Gunderson, L., Holling, C.S. and Walker, B. (2002) 'Resilience and Sustainable Development: Building Adaptive Capacity in a World of Transformations', *AMBIO: A Journal of the Human Environment* 31, 5, 437–440.

Greenhill, B., Mosley, L. and Prakash, A. (2009) 'Trade-Based Diffusion of Labor Rights: A Panel Study, 1986–2002', *American Political Science Review* 103, 4, 669–690.

Hall, P.A. and Soskice, D. (2001) *Varieties of Capitalism. The Institutional Foundations of Comparative Advantage* (Oxford: Oxford University Press).

Hart, S.L. and Dowell, G. (2011) 'A Natural-Resource Based View of the Firm: Fifteen Years After', *Journal of Management* 37, 1464–1479.

Hart, S.L. and Sharma, S. (2004) 'Engaging Fringe Stakeholders for Competitive Imagination', *Academy of Management Executive* 18, 1, 7–18.

Haufler, V. (2001) 'Globalization and Industry-Self-Regulation', in M Kahler and D Lake (eds) *Governance in a Global Economy*, Princeton (NJ: Princeton University Press).

Héritier, A. and Lehmkuhl, D. (2008) (eds) 'The Shadow of Hierarchy and New Modes of Governance', *Journal of Public Policy* 28, 1, 1–17.

Héritier, A., Müller-Debus, A. and Thauer, C. (2009) 'The Firm as an Inspector: Private Ordering and Political Rules, *Business and Politics* 11, 4, Art. 2.

Hoffman, A.J. (1999) 'Institutional Evolution and Change: Environmentalism and the U.S. Chemical Industry', *Academy of Management Journal* 42, 4, 351–371.

Hoffman, A.J. (2001) *From Heresy to Dogma: An Institutional History of Corporate Environmentalism* (Palo Alto, CA: Stanford University Press).

Hoffman, V.H. (2007) 'EU ETS and Investment Decisions: The Case of German Electricity Industry', *European Management Journal* 5, 6, 464–474.

Ikwue, T. and Skea, J. (1994) 'Business and the Genesis of the European Community Carbon Tax Proposal', *Business Strategy and the Environment* 3, 2, 1–10.

IPCC (2007) *The AR4 Synthesis Report: Fourth Assessment Report of the Intergovernmental Panel on Climate Change* (Cambridge, UK and New York, USA: Cambridge University Press).

Kaufman R.R. and Segura-Ubiergo, A. (2001) 'Globalization, Domestic Politics, and Social Spending in Latin America: A Time-Series Cross-Section Analysis, 1973–97', *World Politics* 53, 4, 553–587.

Kolk, A. (2001) 'Multinational Enterprise and Industrial Restructuring: A Strategic Environmental Management Approach', in M Binder, M Jänicke and U Petschow (eds) *Green Industrial Restructuring: International Case Studies and Theoretical Interpretations* (Berlin, Heidelberg and New York: Springer Verlag).

Kolk, A. and Levy, D. (2001) 'Winds of Change: Corporate Strategy, Climate Change and Oil Multinationals', *European Management Journal* 19, 5, 501–509.

Kolk, A. and Pinkse, J. (2005) 'Business Responses to Climate Change: Identifying Emergent Strategies', *California Management Review* 47, 3, 6–20.

Kolk, A. and Pinkse, J. (2008) 'A Perspective on Multinational Enterprise and Climate Change. Learning from the "Inconvenient Truth"', *Journal of International Business Studies* 39, 8, 1359–1378.

Kolk, A. and Pinkse, J. (2009) *International Business and Global Climate Change* (London and New York: Routledge).

Kolk, A. and Pinkse, J. (2012) 'Multinational Enterprises and Climate Change Strategies', in A Verbeke and H Merchant (eds) *Handbook of Research on International Strategic Management* (Cheltenham, UK: Edward Elgar).

Kolk, A., van Tulder, R. and Welters, C. (1999) 'International Codes of Conduct and Corporate Social Responsibility: Can TNCs Regulate themselves?' *Transnational Corporations* 8, 1, 143–180.

Kollman, K. and Prakash, A. (2001) 'Green by Choice? Cross-National Variations in Firms' Responses to EMS-Based Environmental Regimes', *World Politics* 53, April, 399–430.

Kranz, N. (2010) 'What Does It Take? – Engaging Business in Addressing the Water Challenge in South Africa. Governance for Sustainable Development under Weak Regulatory Capacity. PhD Thesis', Freie Universität Berlin, Berlin.

Lazarus, R.J. (2009) 'Super Wicked Problems and Climate Change: Restraining the Present to Liberate the Future', *Cornell Law Review* 94, 1153–1234.

Levy, D. (1997) 'Environmental Management as Political Sustainability', *Organization and Environment* 10, 2, 126–147.

Levy, D. and Kolk, A. (2002) 'Strategic Responses to Global Climate Change: Conflicting Pressures on Multinationals in the Oil Industry', *Business and Politics* 4, 3, 275–300.

Levy, D. and Newell, P.J. (2002) 'Business Strategy and International Environmental Governance: Towards a Neo-Gramscian Synthesis', *Global Environmental Politcs* 2, 4, 84–100.

Levy, D. and Rothenberg, S. (2002) 'Heterogeneity and Change in Environmental Strategy: Technological and Political Responses to Climate Change in the Automobile Industry', in A Hoffman and M Ventresca (eds) *Organizations, Policy and the Natural Environment: Institutional and Strategic Perspectives* (Palo Alto, CA: Stanford University Press).

Liu, J., Dietz, T., Carpenter, S.R., Alberti, M., Folke, C., Moran, E., Pell, A., Deadman, P., Kratz, T., Lubchenko, J., Ostrom, E., Ouyang, Z., Provencher, W., Redman, C.L., Schneider, S.H. and Taylor, W. (2007) 'Complexity of Coupled Human and Natural Systems', *Science* 317, 1513–1516.

Lofdahl, C.L. (2002) *Environmental Impact of Globalization and Trade. A Systems Study* (Cambridge, MA: MIT Press).

Maguire, S., Hardy, C. and Lawrence, T.B. (2004) 'Institutional Entrepreneurship in Emerging Fields: HIV/AIDS Treatment Advocacy in Canada', *Academy of Management Journal* 47, 5, 657–679.

Mayntz, R. and Scharpf, F.W. (1995a) 'Der Ansatz des akteurszentrierten Institutionalismus', in R Mayntz and FW Scharpf (eds) *Gesellschaftliche Selbstregulierung und politische Steuerung* (Frankfurt am Main: Campus).

Mayntz, R. and Scharpf, F.W. (1995b) 'Steuerung und Selbstorganisation in Staatsnahen Sektoren', in R Mayntz and FW Scharpf (eds) *Gesellschaftliche Selbstregulierung und politische Steuerung* (Frankfurt, New York: Campus).

Meyer, J.W. and Rowan, B. (1977) 'Institutionalized Organizations: Formal Structures as Myth and Ceremony', *American Journal of Sociology* 83, 2, 340–363.

Mol, A.P.J. (2001) *Globalization and Environmental Reforms: The Ecological Modernization of the Global Economy* (Cambridge, MA: MIT Press).

Moser, S.C. and Ekstrom, J.A. (2010) 'A Framework to Diagnose Brriers to Climate Change Adaptation', *Proceedings of the National Academy of Sciences (PNAS)* 107, 51, 22026–22031.

Murphy, D. (2000) *The Structure of Regulatory Competition: Corporations and Public Policies in a Global Economy* (Oxford: Oxford University Press).

Nelson, D.R. (2010) 'Adaptation and Resilience: Responding to a Changing Climate', *Wiley Interdisciplinary Reviews Climate Change* 2, 1, 113–120.

Nelson, D.R., Adger, W. and Brown, K. (2007) 'Adaptation to Environmental Change: Contributions of a Resilience Framework', *Annual Review of Environment and Resources* 32, 1, 395–419.

Newell, P. and Paterson, M. (1998) 'Climate for Business: Global Warming, the State and Capital', *Review of International Political Economy* 5, 4, 679–703.

Nidumolu, R., Prahalad, C.K. and Rangaswami, M.R. (2009) 'Why Sustainability Is Now the Key Driver of Innovation', *Harvard Business Review* 7, 1–10.

Parker, C. (2002) *The Open Corporation: Effective Self-Regulation and Democracy* (Cambridge: Cambridge University Press).

Perrow, C. (1985) 'Comment on Lanngton's "Ecology Theory of Bureaucracy"', *Administrative Science Quarterly* 30, 278–283.

Pinkse, J. and Kolk, A. (2009) *International Business and Global Climate Change* (New York: Routledge).

Pinkse, J. and Kolk, A. (2012) 'Addressing the Climate Change – Sustainable Development Nexus: The role of Multi-Stakeholder Partnerships', *Business and Society* 51, 1, 176–210.

Porter, M.E. and Kramer, M.R. (2002) 'The Competitive Advantage of Corporate Philanthropy', *Harvard Business Review* 80, 12, 56–68.

Porter, M.E. and Kramer, M.R. (2011) 'Creating Shared Value: How to Reinvent Capitalism – and Unleash a Wave of Innovation and Growth', *Harvard Business Review* 89, 1–2, 62–77.

Porter, M.E. and van der Linde, C. (1995) 'Toward a New Conception of the Environment-Competitiveness Relationship', *Journal of Economic Perspectives* 9, 4, 97–118.

Potoski, M. and Prakash, A. (2005) 'Green Clubs and Voluntary Governance: ISO 14001 and Firms' Regulatory Compliance', *American Journal of Political Science* 49, 2, 235–248.

Potoski, M. and Prakash, A. (2006) 'Racing to the Bottom? Trade, Environmental Governance, and ISO 14001', *American Journal of Political Science* 50, 2, 350–364.

Potoski, M. and Prakash, A. (2009) *Voluntary Programs: A Club Theory Perspective* (Cambridge MA: MIT Press).

Prakash, A. and Potoski, M. (2006) *The Voluntary Environmentalists: Green Clubs, ISO 14001 and Voluntary Environmental Regulations* (Cambridge: Cambridge University Press).

Prakash, A. and Potoski, M. (2007) 'Investing Up: FDI and the Cross-Country Diffusion of ISO 14001 Management Systems', *International Studies Quarterly* 51, 3, 723–744.

Risse, T. (2011) 'Governance in Areas of Limited Statehood: Introduction and Overview', in T Risse (ed.) *Governance without a State? Policies and Politics in Areas of Limited Statehood* (New York: Columbia University Press).

Ruggie, J.G. (2004a) 'How to Marry Civic Politics and Private Governance', in CCoEaI Affairs (ed.) *The Impact of Global Corporations on Global Governance* (New York: Carnegie Council on Ethics and International Affairs).

Ruggie, J.G. (2004b) 'Reconstituting the Global Public Domain: Issues, Actors and Practices', *European Journal of International Relations* 10, 4, 499–532.

Scharpf, F.W. (1997) *Games Real Actors Play. Actor-Centered Institutionalism in Policy Research* (Boulder, CO: Westview Press).

Skjaerseth, J.B. and Skodvin, T. (2003) *Climate Change and the Oil Industry: Common Problems, Varying Strategies* (Manchester: Manchester University Press).

Spar, D.L. and LaMure, L.T. (2003) 'The Power of Activism: Assessing the Impact of NGOs on Global Business', *California Management Review* 45, 78–101.

Sullivan, R. (2009) *Corporate Responses to Climate Change: Achieving Emissions Reductions through Regulation: Self-Regulation and Economic Incentives* (Sheffield, UK: Greenleaf).

Swart, R., Robinson, J. and Cohen, S. (2003) 'Climate Change and Sustainable Development: Expanding the Options', *Climate Policy* 3, Supplement 1: S19–S40.

Van den Bergh, J.C.J.M., Truffer, B. and Kallis, G. (2011) Environmental Innovation and Societal Transitions: Introduction and Overview, *Environmental Innovation and Societal Transitions* 1, 1, 1–23.

Vogel, D. and Kagan, R. (2004) (eds) *Dynamics of Regulatory Change: How Globalization Affects National Regulatory Policies* (Berkeley, Los Angeles: University of California Press).

Walker, B., Holling, C.S., Carpenter, S.R. and Kinzig, A. (2004) 'Resileinnce, Adaptability and Transformability in Social Ecological Systems', *Ecology and Society* 9, 2, 5.

Xing, Y. and Kolstad, C. (2002) 'Do Lax Environmental Regulations Attract Foreign Investment?', *Environmental and Resource Economics* 21, 1, 1–22.

Ziervogel, G. and Taylor, A. (2008) 'Feeling Stressed: Integrating Climate Adaptation with Other Priorities in South Africa', *Environment: Science and Policy for Sustainable Development* 50, 2, 32–41.

Zucker, L.G. (1977) 'The Role of Institutionalization in Cultural Persistence', *American Sociological Review* 42, 726–743.

2
Corporate Response to Climate Change in Areas of Limited Statehood: An Outline of the Organisational Configurations in Kenya and South Africa

Farai Kapfudzaruwa

Introduction

Climate change has attracted business attention because of its actual and potential strategic impact on companies. It presents risks and opportunities for those companies that produce fossil fuels (such as oil or utilities companies), those that depend on these fuels (e.g. chemical companies and airlines) and those that want to develop new market opportunities arising from risks coverage or emerging emission trading systems (e.g. banks, insurance companies) (Kolk and Pinkse, 2004). On the whole, as a result of climate change, corporations will have to respond to government regulators and stockholders, protect and enhance their ethical images, avoid legal liabilities and develop new business opportunities, in order to remain competitive as changes in prices, technology and demand patterns disrupt both sectors and entire supply chains (Berry and Rondinelli, 1998; Ozawa-Meida et al., 2009). However, firms operating in developing countries, such as South Africa and Kenya, are faced with numerous complexities, due to these countries' 'limited statehood'. Risse and Lehmkuhl (2006: 9) define 'limited statehood' as 'deficits by a nation state to perform its core functions of monopolising the use of force and ability to enforce political decisions'. This chapter explores corporate climate change strategies in countries that have varying degrees of deficits in their abilities to steer effective climate mitigation and adaptation. It provides an overview of how companies are responding to climate change in Kenya and South Africa, two countries with varying levels of 'limited statehood'.

We expect that firms operating in South Africa, which is a 'newly industrialising country' with emerging climate change policies but gaps in enforcement, are more responsive to climate change than firms operating in Kenya, which is a 'developing country' with more prominent gaps in its climate change policies. South Africa's National Climate Change Response Policy was approved by cabinet in 2011 and provides a clear roadmap for the country's response to climate change. In addition, in its 2012/13 budget the treasury announced that it will introduce a carbon tax of R120 (US$14) per ton of CO_2e for emissions above set thresholds. The tax will come into effect in 2013/14 and will increase by 10 per cent a year until 2020. Furthermore, South Africa has a string of environmental regulations for air pollution. However, despite these fairly strong emerging policies, South Africa still has a weak administrative capacity for implementing regulations and securing compliance (Börzel et al., 2011; Hönke et al., 2008).

Kenya, in contrast, does not have any explicit climate change regulations or policies. The National Climate Change Response Strategy (NCCRS), which was penned in 2010, has not been developed into any meaningful policy. These efforts are being undermined by an intense socio-economic growth orientation in government regulations and policies, which has resulted in misaligned incentives and priorities for business. This is worsened by an uncoordinated bureaucratic system, with inherent conflicts between the fragmented agencies in government (CSR consultant, personal communication, May 2011).

Despite its challenges, the government in South Africa is capable of guiding companies on the use of appropriate strategies to manage their greenhouse gas (GHG) emissions. The emerging policies can also be used by government to cast a 'shadow of hierarchy', in order to provide incentives to the private sector to engage government in non-hierarchical coordination in climate change governance. In Kenya, the domination of socio-economic policies over the barely visible climate change policies, together with the fragmented government agencies, deprive the government of the ability to enforce or engage the private sector in managing their GHG emissions.

In addition to varying levels of limited statehood, companies' responses to climate change are also expected to vary across as well as within sectors in these two countries. This is because of the different sectoral institutions and issue salience in these sectors. Generally, energy-intensive companies, which face more physical, regulative and economic risks from climate change, tend to be associated with institutions such as the industry associations that guide them to be responsive to climate change.

Furthermore, some companies possess specific organisational capabilities, which enable them to be innovative and to tackle climate change.

To explore how and why companies in Kenya and South Africa respond to climate change, this study analysed the content of websites, sustainability reports and annual reports of companies listed on the Nairobi Stock Exchange (NSE) in Kenya and the Johannesburg Stock Exchange (JSE) in South Africa, supported by structured and unstructured interviews. Using cluster analysis, we identify the main organisational configurations in relation to climate change responses. The results from the study reveal four clusters in South Africa (laggards, emergent planners, efficiency drivers and visionaries) while in Kenya three clusters are identified (laggards, emergent planners and efficiency drivers). An analysis of these results reveals that performance with regard to climate is very variable, with most companies, particularly in Kenya, being low performers (that is, laggards). In addition to this, the clusters are horizontal, indicating that few companies perform in different dimensions, for example the 'visionaries' consistently performed well on all indicators while the 'laggards' consistently performed poorly on all indicators. The results of the cluster analysis also reveal variability within the country, with regard to the corporate climate change responses. This is due to sector-specific institutions and corresponding perceived issue salience, along with different company-specific organisational resources and culture. Finally, the analysis also shows variability in responses between the countries, with South African companies being more responsive than Kenyan ones, because of the varying levels of limited statehood. In order to provide a background and to explain the theoretical concepts used in exploring the results from the cluster analysis, this chapter first discusses corporate climate change responses in Africa and the typology used to develop the configurations.

Business and climate change in Africa

It has been argued that the private sector in Africa is well placed to make a significant contribution to improving social and environmental conditions in Africa (Visser, 2005; for a more cautious view, see Hamann, 2006). Business is asked to play a prominent role in efforts to mitigate, and adapt to, the effects of climate change, because Africa is one of the regions that have been identified as particularly vulnerable to the impacts of climate change (UNFCC, 2007; Vogel, 2009). At the same time, there is an urgent need for the private sector – as investors, innovators, manufacturers and employers – to lead efforts in reducing poverty and unemployment. Africa is the continent where social needs are greatest. Life expectancy in many parts of Africa within the low human development index is still only 51 years on average (and as low as 38 years in some countries), while gross income per capita averages US$862 (dropping as low as US$90 in some places) and adult literacy is less than 48 per cent in some countries (UNDP, 2009). Despite being relatively well-endowed with energy resources, Africa generates only 3.1 per cent of the world's electricity, and this is a major barrier to economic development (UN, 2009). Energy production tends to be costly, relying heavily on fossil fuels (about 80 per cent of electricity generation) (UN, 2009).

In addition to these challenges, Africa presents one of the most complex operating environments for the private sector, as a result of the varying levels of 'limited statehood' among the few 'new industrialising economies', such as South Africa and Morocco, and the many 'weak' states in transition (Uganda, Tanzania), together with the 'failed or failing states' such as Congo and Zimbabwe (Risse and Lehmkuhl, 2006). As outlined above, South Africa is generally considered a continental leader in climate change, even though it still lacks capacities to monitor and enforce state regulation (Borzel et al., 2011). As a result, many of the companies operating in South Africa that are directly affected by climate change are responsive it because they are anticipating the effects of impending regulations such as the carbon tax. In contrast, many African countries do not have climate change specific legislation and policies. Most of them have broad, fragmented environmental legislation, which still does not hold business firms accountable (Vogel, 2009).

In addition to the country variation in responding to climate change, there is also sectoral variation. Generally, we would expect energy-intensive sectors to face more direct regulative, physical and economic risks than those sectors that are less energy intensive (Kolk and Pinkse, 2009). These energy-intensive sectors are also likely to be more responsive to concerns about energy consumption and climate change. Furthermore, these responsive sectors are surrounded by institutions that influence them in a way that makes them more responsive to climate change (for example, industry associations) (see Greenwood et al., 2002). In South Africa, companies in mining and other energy-intensive sectors, which consume 40 per cent of the country's energy, have formed the Energy Intensive User Group (EIUG) to assist them to respond collectively to climate change and other energy related issues (see Chapter 5). Within specific sectors, companies might have varied responses to climate change because of their different organisational capabilities and culture (Hoffman, 2010). For example, some companies are endowed with leadership that recognises climate change as a grand challenge, which provides risks and opportunities to the company. This filters through the whole company and drives innovation towards climate change mitigation adaptation (Pervaiz, 1996).

Research approach and methodology

This study is based on a content analysis of annual reports, sustainability reports and web-based content (all of which will be referred to as 'reports') of the 45 listed companies on the NSE and the top 100 companies listed on the JSE. The sectoral composition for the Nairobi Stock Exchange (NSE) and the Johannesburg Stock Exchange (JSE) samples are illustrated in Figures 2.1 and 2.2, respectively (see Chapter 1). This sample is considered representative of listed African firms, because it is drawn from two prominent stock exchanges from two different regions (eastern and southern Africa, respectively), and it includes companies of diverse size, economic scope and geographic reach. To illustrate, the sample ranges from Rea Vipings Plantations, a Kenyan agriculture company with a turnover of US$18 million, to BHP Billiton, one of the largest resources companies in the world, with a turnover of over US$50 billion.

The variables for the data were generated using a selection of climate change indicators linked to the six strategic options presented in the typology by Kolk and Pinkse (2005) in the introduction (see Figure 1.2 in Chapter 1). Organisational configurations can be defined as commonly occurring clusters of attributes of organisational strategies, structures and processes (Mintzberg, 1983; Miles and Snow, 1978; Miller, 1987). The intention of the configurational approach is to increase the understanding of organisational phenomena by identifying a distinct and internally consistent set of firms (Ketchen et al., 1993). As a result, configurations have been used prominently in exploring the determinants of performance. Similarly, improved understanding of corporate responses to climate change can be achieved by identifying distinct and coherent sets of firms.

Kolk and Pinkse's (2005) typology of companies' strategic responses to climate change (Figure 1.2, Chapter 1) focuses on mitigation only. According to the typology, companies' responses to climate change focus on two overarching strategic aims: *innovation* and *compensation*. When these two overarching strategic aims are combined with different levels of organisational activities and interactions, a matrix is developed to outline the strategic options in response to climate change (Kolk and Pinkse, 2005). In the resulting typology, six strategic options emerge

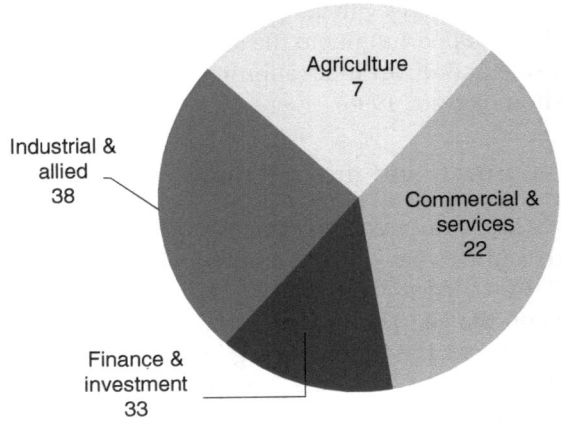

Figure 2.1 Sectoral representation of sample companies on the Nairobi Stock Exchange (Kenya)*
* The categories are based on the company listings on the stock exchange.

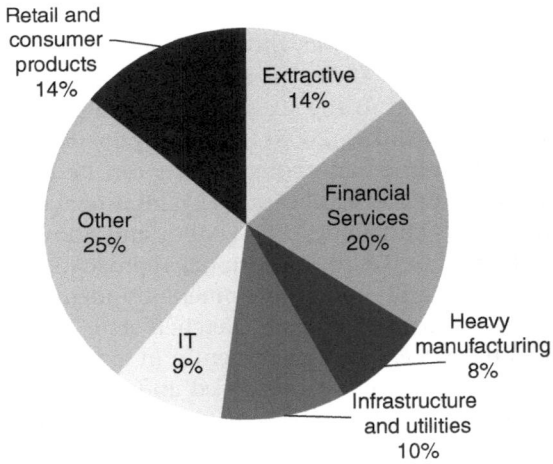

Figure 2.2 Sectoral representation of sample companies on the Johannesburg Stock Exchange (South Africa)**
** The categories are based on the company listings on the stock exchange.

that could be part of a comprehensive strategy for climate change in which companies combine several options (op cit.). The vertical axis distinguishes the three different levels of organisational activities and interactions: within the individual company (*internal*), within the value chain (*vertical*), and with other private or public actors (*horizontal*). The horizontal axis outlines the two main drivers (innovation and compensation) of corporate responses to climate change.

Table 2.1 outlines a combination of the six strategic options in the Kolk and Pinkse typology and the indicators used in this study. The categories in Table 2.1 are related to the main strategic aims of *innovation and compensation* in the typology that have a strong emphasis on climate change mitigation. Organisations that have a strong internal focus through improving their dynamic capabilities and view climate change as an opportunity will tend to emphasise *innovation* in their GHG management. This means they will focus on improving their environmental technologies and services to reduce their emissions As a result, these companies will focus on *process improvement, product development* and *new product and market combinations*. On the other hand, there are companies that do not have the technical and organisational capabilities to innovate. Instead, these companies will focus on *compensation* by borrowing emission reduction technologies from other companies through *GHG accounting and internal transfers, supply chain measures* and *acquisition of emission credits* or *political activism*.

The first category in *process improvement* focuses on understanding the degree to which the company is developing its resources and capabilities to tackle climate change through different internal processes. Hence, the indicators are meant to identify the level of executive commitment to enhancing process efficiencies in the company to reduce GHG emissions, and whether staff have concrete incentives to do this. In addition, the category includes indicators measuring the specific investment in equipment and resources to reduce emissions, and whether such initiatives are making a difference. The category in *GHG accounting and internal transfers* provides indicators that are meant to measure the internal compensatory measures within the company. These indicators are meant to show the extent to which the company measures and discloses its GHG emissions and targets, and plans to internally transfer these emissions to other companies or business units in different locations.

The *product development* and *supply chain measures* categories comprise indicators exploring the companies' supply chain responses to climate change. The *product development* indicators explore a company's innovation initiatives to develop new products and services within particular industries and the review of progress and status of these products and services in reducing GHG emissions. The *supply chain measures* indicators focus on the compensatory activities of the companies within

Table 2.1 An outline of the strategic options and indicators used in the content analysis

Organisational level	Strategic aim	
	Innovation	Compensation
Internal (company)	*Process improvement*	*GHG Accounting and Internal Transfers*
	• The company builds and develops resources to enhance its process efficiencies	• The company measures and discloses its scope 1, 2 and 3 GHG emissions
	• The company invests in new equipment and initiatives to enhance process efficiencies and resource productivity to reduce its GHG emissions	• The company has specific internal emission reduction targets over a certain period of time
	• The company has an executive board or committee responsibility to align the company's goals and its process improvement initiatives to reduce GHG emissions	• The company has an internal emission reduction plan to achieve the set targets
	• The company provides incentive mechanisms for individual management of climate change issues	• The company has a pilot project(s) to internally transfer its emissions
	• The company mentions concrete results of process improvement initiatives that have been achieved	• The company has an operational internal emission transfer scheme
Vertical (supply chain)	*Product development*	*Supply chain measures*
	• The company has a product innovation policy to develop climate-friendly products and services to respond to climate change	• The company outsources its GHG intensive activities (e.g. transportation)
	• The company participates in the development of climate-friendly products and services	• The company procures its electricity from renewable energy sources

(continued)

Table 2.1 Continued

Organisational level	Strategic aim	
	Innovation	Compensation
	• The company has mechanisms to review its progress and status in the development of climate-friendly products and services	• The company sets emission reduction targets for its supply chain
	• The company participates in marketing of green/ climate-friendly products and services as part of its core business	• The company evaluates the supply chain's emission reduction performance using industry standards and certification (ISO 14001)
	• The company accounts for the GHG emissions reduced from its climate-friendly products and services	• The company is a member of a professional/industrial association which influences its response to climate change
Horizontal (beyond supply chain)	*New product and market combinations*	*Acquisition of emission credits and political activity*
	• The company explores the possibility of partnerships in developing climate-friendly products and entrance of new markets	• The company participates in carbon offset projects (e.g. CDM)
	• The company has a policy and targets to participate in cross-sectoral collaborations to develop new product and market combinations	• The company reports carbon credits or allowances traded
	• The company has concrete partnerships with private sector partners to develop new climate-friendly product and market combinations (specify)	• The company responds to different regulations, norms and cognitive forces related to climate change
	• The company has concrete partnerships with public sector partners to develop new climate-friendly products and entering new markets	• The company has explicit political strategies/activities to influence climate change policy (e.g. lobbying)

(*continued*)

Table 2.1   Continued

Organisational level	Strategic aim	
	Innovation	Compensation
	• The company has concrete partnerships with civil society partners to develop new climate-friendly products and entering new markets	• The company negotiates and collaborates with its cross-sectoral partners (e.g. government, civil society) to shape climate change policy
	• The company has concrete partnerships with the private sector, public sector and civil society partners to develop new climate-friendly products and entering new markets	• The company has explicit strategies to influence its stakeholders' views on climate change.

Adapted from Kolk and Pinkse, 2005.

their supply chain. The indicators look at the role of cognitive and normative pressures, such as standards in a company's climate change strategy. Specifically, these indicators are meant to show the extent to which the company transfers its emissions within its industry, either in response to pressure from other companies or through industry associations. The *new product and market combinations* and *acquisition of emission credits and political activity* categories focus on the climate change responses of the companies through cross-sectoral approaches. The *new product and market combinations* category focuses on how the company evaluates and influences its supply chain to develop new products and enter new markets, by developing and marketing climate-friendly products and services in different sectors. In addition, the category explores the role of different stakeholders in influencing how the company develops new product and market combinations. Finally, the *acquisition of emission credits and political activity* category is aimed at understanding how the company responds to the regulations, surrounding norms and cognitive forces, its level of participation in carbon offset projects, and the extent to which it collaborates and negotiates with a variety of stakeholders from different sectors and backgrounds in influencing and shaping climate change policy.

Each company was given a score between 0 and 3, depending on how systematically and rigorously the indicators derived from the categories

Table 2.2 Scoring system used in the content analysis

Score	Criteria
0	No information about the indicator is provided
1	Basic information relevant to the indicator is provided, but there is no link to company strategy or operations
2	Information is provided about the indicator, including basic information about strategic intent and operational aspects
3	Comprehensive information is provided on the company's approach to that indicator, including strategic intent, implementation and monitoring

were reported. The scoring schematic is provided in Table 2.2. The scores are a reflection of the company's diligence in public reporting on their climate change policies. The key methodological requirement of the scoring schematic is that there is relative consistency between the scores for different indicators and for different companies (see Hamann et al., 2009). To enhance the reliability of the scoring process, two scorers carried out the process. The first scorer carried out the scoring for all the companies and the second scorer conducted the scoring for 20 per cent of the sample using the scoring system used in Table 2.2. These two sets of scores were compared to ensure inter-temporal and inter-scorer reliability (Lombard et al., 2002). The Krippendorff alpha (Hayes and Krippendorff, 2007), used to measure the degree of agreement between these scoring sets, was 0.81, representing sufficient reliability (Krippendorff, 1980, 2004).

It is apparent that data derived from the content analysis is based on company disclosure, not actual policies (their output). This implies that the reliance on disclosure means that some important initiatives by companies might be missed. In an effort to counter this inherent weakness, 55 structured and unstructured interviews were carried out on 18 companies in both countries in the banking and finance sector, food and drink manufacturing sector and industrials and allied sector. At least one or more interviews were carried out for each company in both countries. However, in some companies, particularly in South Africa, more of the interviews were of individuals who were well informed about the company's climate change strategies. Furthermore, interviews were carried out with consultants and individuals knowledgeable about corporate climate change strategies in both countries. The interviews furthered understanding of the company's actual performance in response to climate change and drivers of these strategies. The interviews were carried out with individuals who have considerable knowledge about the climate change initiatives of the case study companies (environmental and CSR managers, operations, supply chain, production managers, consultants and academics). In South Africa, 33 interviews lasting between one and two hours were conducted and in Kenya 22 interviews were conducted.

SPSS was used for the cluster analysis, which was designed to identify individually the different organisational configurations. Clustering is more appropriate than factor analysis and multidimensional scaling in content analysis because it is based on 'intuitively meaningful similarities among units and its resulting hierarchies resemble the conceptualisation of text on various levels of abstraction' (Krippendorff, 2004: 210). For the cluster analysis, the data consists of the categories of strategic options discussed above for each of the 145 firms (see Figure 1.2 in Chapter 1). For each category, for example process improvement, the firms are rated between 1 and 3 on several questions (indicators). Therefore, we computed the average response for each category for each firm to configure their climate strategies.

Results and conclusions

The results of the analysis are shown in Figures 2.3, 2.4 and 2.5, which summarise the mean values of the final cluster centres for the six categories used (Table 2.1). The results show that four different strategy configurations for climate change can be identified in South Africa (*visionaries, efficiency drivers, emergent planners* and *laggards*), while in Kenya three clusters emerged (*internal explorers, emergent planners* and *laggards*) (see Table 2.3). The '*visionaries*' are the companies who have understood the risks and opportunities presented by climate change. Therefore, they are proactively adopting strategies internally, within the supply chain and through collaboration with relevant stakeholders. The '*efficiency drivers*' have a strong internal focus as they start to seriously respond to climate change. The '*emergent planners*' are those companies that are starting to explore different climate change strategies, while the '*laggards*' do not respond to climate change or adopt cosmetic initiatives to respond to climate change.

Even though the clusters in both countries use the same labels for the three identical clusters in the country specific analysis, they cannot be compared directly because, as illustrated in Figures 2.4 and 2.5, the characteristics of the clusters are often not the same in the two countries.

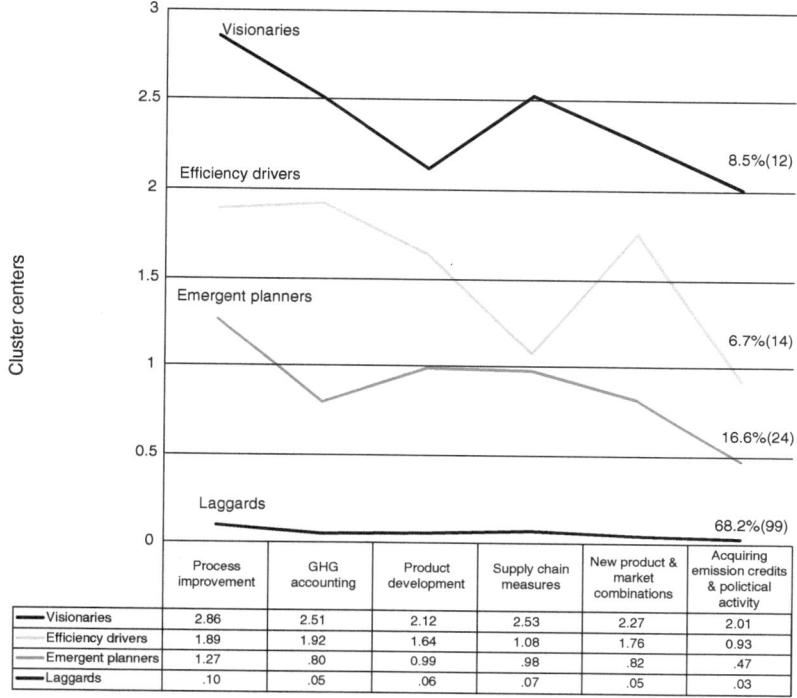

	Process improvement	GHG accounting	Product development	Supply chain measures	New product & market combinations	Acquiring emission credits & polictical activity
Visionaries	2.86	2.51	2.12	2.53	2.27	2.01
Efficiency drivers	1.89	1.92	1.64	1.08	1.76	0.93
Emergent planners	1.27	.80	0.99	.98	.82	.47
Laggards	.10	.05	.06	.07	.05	.03

Figure 2.3   Cluster centres for strategic climate change configurations in South Africa and Kenya

In addition, the Kenyan sample dominates the companies that are non-responsive to climate change, while the South African sample dominates the responsive companies. This suggests that, in addition to other organisational and institutional factors discussed below, South African companies are gradually starting to respond to the slow introduction of policies that address threats of climate change in the country. On the other hand, Kenya does not have explicit climate change policies to compel companies to start to reduce their GHG emissions.

In addition to these country differences, the results also reveal in-country differences along sectors. Due to the varying sectoral institutions and issue salience, the energy-intensive sector in South Africa, for example, is more responsive to climate change than the banking sector because climate change directly affects their operations and manufacturing companies are surrounded by institutions within their industry

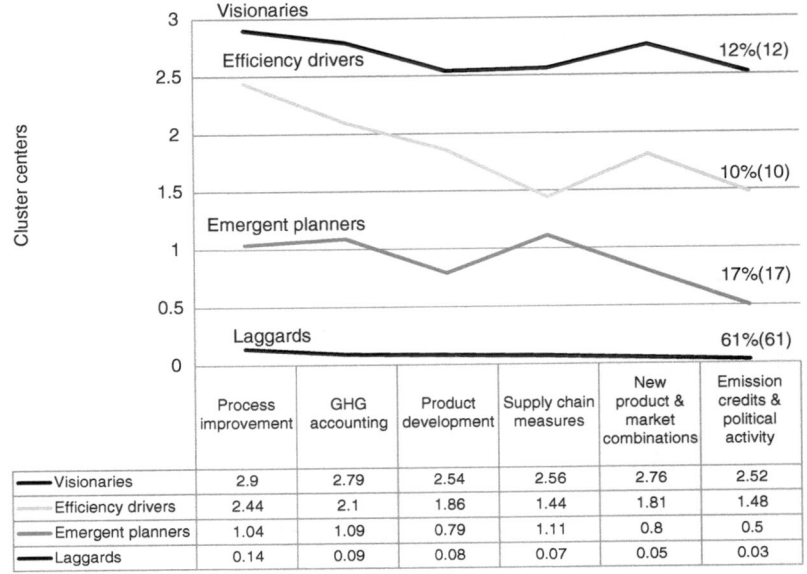

	Process improvement	GHG accounting	Product development	Supply chain measures	New product & market combinations	Emission credits & political activity
—— Visionaries	2.9	2.79	2.54	2.56	2.76	2.52
Efficiency drivers	2.44	2.1	1.86	1.44	1.81	1.48
—— Emergent planners	1.04	1.09	0.79	1.11	0.8	0.5
—— Laggards	0.14	0.09	0.08	0.07	0.05	0.03

Figure 2.4 Cluster centres for strategic climate change configurations in South Africa

associations, which will tend to influence them to become energy efficient. Finally, within the sectors, certain companies are more responsive to climate change than others. For example, in the retail sector, Woolworths is an industry leader with its climate change mitigation and adaptation strategies, because of company specific capabilities and culture which are less prominent in other retailers. In the following, each of the clusters will be discussed in more detail.

Laggards

The '*laggards*' represent the largest proportion (68 per cent) of companies in both countries, indicating that a large proportion of the private sector, particularly in Kenya is still struggling to participate in climate change governance. As shown in Figure 2.4 and 2.5, a larger proportion (84 per cent) of the companies sampled in Kenya belongs to this cluster compared to the South African sample (61 per cent). The companies in this cluster have very low scores (below 0.5) on all the indicators in both countries, and they therefore share the

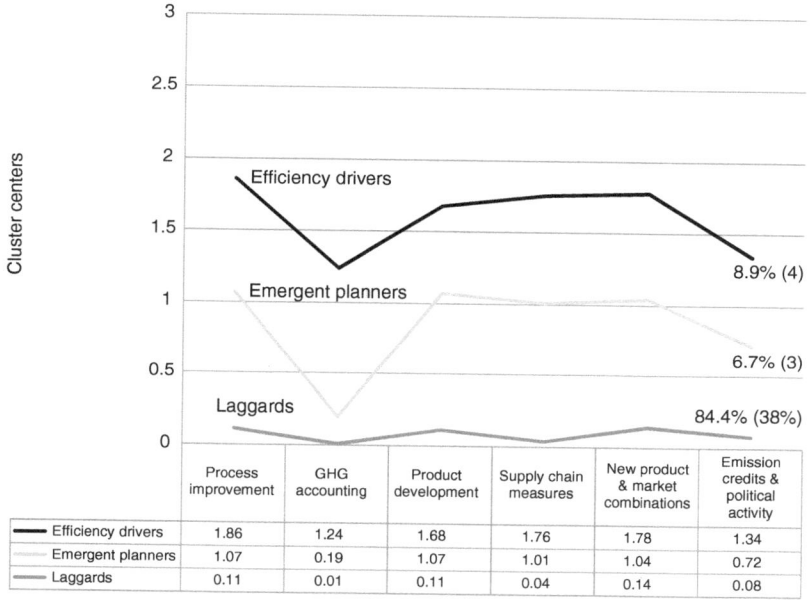

	Process improvement	GHG accounting	Product development	Supply chain measures	New product & market combinations	Emission credits & political activity
▬▬ Efficiency drivers	1.86	1.24	1.68	1.76	1.78	1.34
Emergent planners	1.07	0.19	1.07	1.01	1.04	0.72
▬▬ Laggards	0.11	0.01	0.11	0.04	0.14	0.08

Figure 2.5   Cluster centres for strategic climate change configurations in Kenya

same attributes. These very low scores indicate that the companies do not have any concrete internal strategies nor are they collaborating with any of their sectoral partners in responding to climate change. The financial and services and information and technology industries particularly in Kenya, dominate this cluster. Companies in this cluster include, for example, Kenya Commercial Bank, Access Kenya, Datatec (Kenya) and Tiger Brands in South Africa (see Table 2.1 for a summary of the companies' strategies). Many Kenyan companies, particularly those in the commercial and services sector, focus on planting trees as their main response to climate change, partly because it requires less investment of resources and innovation. While afforestation assists in increasing the carbon sink, which absorbs GHGs, it does not help the companies reduce their direct scope 1 (direct) and 2 (indirect) emissions, Hence they had low scores in all the indicators.

Intensively pro-growth policies, which do not adequately integrate climate change, particularly Vision 2030 which is the blueprint of Kenya's future, are sending 'misaligned incentives to a majority of

Table 2.3 Summary of strategies adopted by example companies representing the corresponding clusters

Cluster	Company	Strategy
Laggards	Access Kenya	The internet and technology company reports its 'continued excellence, leadership and stewardship in tackling climate change', but there is no evidence in their reporting that they measure their GHG emissions, forecasted future emission or energy use to develop a mitigation plan or any of the strategic options in Table 2.1
	ABSA bank	The bank does not have a notable strategy to respond to climate change. According to a risk analyst at the bank, 'there is not enough data on how climate change will affect their corporate clients in terms of risks to the bank to enable them to make important decisions' (ABSA Capital Risk Analyst, interview).
Emergent Planners	BOC Kenya	BOC Kenya, a supplier of industrial, process and specialty gases has outlined plans to adopt climate change strategies from its parent company, the Linde Group. The plans include capitalising on the rise in demand for large-scale thin film photovoltaic modules by supplying specialty gases 'that help keep production as climate neutral and cost effective as possible'. The Linde Group has been involved in 'the evolution of hydrogen technology covering the entire hydrogen value chain from generation and liquefaction through transport solutions to vehicle fuelling' (Linde Group annual report, 2010).
	Group Five	This construction company has been participating in the South African Carbon Disclosure Project since 2009, which allowed it to measure its GHG emissions using the Greenhouse Gas Protocol. Therefore, it is in the process of setting targets based on the data from Green Gas Protocol. In an effort to reach these targets, the company has started being involved in green buildings and renewable energy. However, the company acknowledges that 'it still has to develop in-house skills and a comprehensive strategy to accommodate climate change' (Group Five, SHE Officer, interview)
Efficiency Drivers	Mumias Sugar	Mumias Sugar, a Kenyan milling company, has been implementing energy efficiency initiatives through the Energy Efficiency Accord championed by the Kenya Association of Manufacturers' CEEC unit. Through its energy efficiency project, the company has identified and corrected energy consumption levels using 'medium voltage variable frequency drives' and modified its fuel and fleet management system. In addition, the company recently completed its bagasse cogeneration plant, which has enables the company to supply 26 megawatts of renewable energy to the national grid. As result of cogeneration, the company does not have to rely on energy sources such as coal, which emit large amounts of GHGs.

Sappi	The global pulp and paper company based in South Africa has been carrying out numerous climate change mitigation efforts which earned them a score of 75% in the South Africa Disclosure Leadership Index of the CDP in 2010. Through production and energy efficiency programmes the company has managed to reduce its CO_2 emissions by 25% over the past five years. Sappi is currently generating 38% of its electricity from renewable sources in South Africa through its cogeneration plants using black liquor, bark, sludge and biomass from its operations
Visionaries Woolworths	South African retailer Woolworths launched its first CO_2 refrigerant pilot store, together with an eco-fridge truck refrigeration, in 2010, as part of its plans to reach its target to reduce its carbon footprint by 30% by 2012. Adaptation strategies such as sustainable farming practices through the Farming for the Future programme have enabled the retailer to reduce its farmers' water consumption. In addition, the company is driving innovation in clothing by selling a range products made from bamboo, a plant that doesn't use fertilisers, helps improve soil quality and absorbs moisture three times better than cotton (see Methner chapter in this volume).

companies operating in Kenya' (CSR consultant, personal communication, May, 2011). As a result, many companies, especially the less energy-intensive ones in the financial and services and information technology industries (which represent most companies in the Kenya sample) do not have incentives to tackle climate change. Moreover, the alignment challenges within the bureaucracy, resulting in inherent conflicts between the fragmented government agencies, and the absence of external pressure from different actors such as civil society, results in inaction on climate change issues (Marquis et al, 2012).

In South Africa, through the White Paper on the National Climate Change Response Strategy and the introduction of carbon tax in the 2013/14 financial year by treasury, the state is showing its ability to enforce binding decisions on the private sector. However, most of these initiatives target the energy-intensive companies. This means that the less polluting industries, particularly the financial services, IT and consumer services which constitute the majority of companies in this cluster, have little motivation to respond to climate change. This is shown by the absence of many South African financial services firms in the carbon trading markets such as the Clean Development Mechanisms (CDM). Companies in this cluster such as Access Kenya and Tiger Brands, which recognise climate change as an emerging strategic issue for them, avoid implementing any concrete strategies to respond to the issue, but rather adopt cosmetic initiatives such as tree planting.

Emergent planners

This cluster consists of companies that have recognised the risks and opportunities that climate change presents to their operations. Therefore, they have put targets in place and are in the early phases of comprehensively implementing strategies to meet these targets. In both countries, companies in this cluster had a moderately low score, between 1.3 and 0.5, which indicates that they are in the initial phases of tackling climate change. However, in Kenya the cluster has a very low score of 0.19 on *GHG accounting and transfers*. As is the case with laggards, many of these Kenyan companies in this cluster do not emit a large amount of GHG emissions, therefore, 'they feel no need to measure their emissions' (Risk Manager, Cooperative Bank of Kenya, personal communication, May, 2011).

The 'emergent planners' in both countries are represented by a mixture of companies from different sectors. However, South African companies dominate the composition of this cluster (17 per cent) compared with

Kenyan companies (6.7 per cent). This reflects the general trend of being responsive to climate change among many South African companies in this sample. Like the South African *'efficiency drivers'* and *'visionaries'*, the *'emergent planners'* are reacting to the threat and potential impact of the proposed carbon tax in 2013/14 by the treasury in South Africa, and the anticipated regulations emanating from the White Paper on the National Climate Change Response Strategy, which was approved by cabinet as policy in November 2011. 'Companies will need respond to the risks posed by the carbon tax because, according to our research, if the FTSE/JSE top 40 companies were to pay the proposed carbon tax of R120 (US$ 14) per ton of CO_2e per tonne of CO_2 for all their direct operational emissions, their carbon costs could amount to almost US$974 million' (sustainable investment architect, Sinco, personal communication, January, 2012).

While Kenya does not emit large amounts of GHGs, scientists are forecasting that the country will be one of the countries most affected by the effects of climate change, particularly the agriculture sector (Kabubo-Mariara and Karanja, 2007). Therefore, companies such as BOC Kenya (see Table 2.2) and Kakudzi and Rea Vipings Plantations representing the 'emergent planners' will be expected to be responsive to the forecasted shortages of water and high temperatures in the Rift Valley. These companies are reportedly recruiting an analyst from Europe to help set targets and plan for these forecasted scenarios (Maggie Opondo, University of Nairobi academic, personal communication June, 2011).

Efficiency drivers

This cluster encompasses companies that are starting to have climate strategies with a strong internal focus. Therefore, they scored significantly higher on indicators that are driven by resource efficiencies, innovation, managerial capabilities and the structure and culture within the company. However, the attributes of the cluster in the two countries are slightly different. First, the South African companies scored significantly higher than Kenyan companies on *process improvement* and *GHG accounting and transfers*. This is partly because many of the Kenyan efficiency drivers' energy efficiency initiatives are motivated by energy security (see Chapter 5). Therefore, they tend not to measure their GHG emissions. South African companies scored higher in this cluster partly because the Energy Efficiency Accord has been running for longer (since 2005) in South Africa compared to Kenya, where it was launched in November 2011. Therefore, the South African efficiency drivers have been implementing these process improvement strategies for longer (see Chapter 5).

The four Kenyan efficiency drivers are active members of the Kenya Association of Manufacturers (KAM) Centre for Energy Efficiency and Conservation. The centre has launched an energy efficiency programme for its members, but the initiative is still in its early stages and has been beset by funding constraints (KAM official, personal communications, November 2010). As a result, while the Kenyan 'efficiency drivers' have started measuring their emissions and attempting to improve their energy efficiencies they still face many challenges.

Many of the 'efficiency drivers' are still focusing on their in-house strategies, by measuring their GHG emissions and focusing on energy and resource efficiency. This means that they do not have a strong focus on other activities within and beyond the supply chain. As a result, both Kenyan and South African samples have average scores on *product development* and *product and market combinations* indicators (1.67–1.87). Furthermore, this cluster performed poorly on compensatory activities in collaboration stakeholders within or beyond the supply chain, resulting in low scores on *supply chain measures* and *acquiring emission credits and political activity*.

Visionaries

The 'visionaries' cluster, which is only present in South Africa, represents companies that are explicitly emphasising and addressing the opportunities and risks presented by climate change. They therefore scored fairly well on all the indicators (Figure 2.4). Woolworths, which had a score of 83 per cent in the CDP Leadership Index, is a good example of a visionary (see Table 2.2). Like many visionary companies, Woolworths's leadership has committed financial and technical resources to their 'Good Business Journey', which aims to improve the company's sustainability performance. The company has been able to set climate change targets and develop a management system to monitor the progress different departments are making to meet the targets (see Methner, Chapter 7). As a result, the commitment to climate change by the company's leadership has been transmitted to the whole company. Moreover, many of the visionaries' climate change strategies are influenced by a certain group of stakeholders directly linked to the company's operations. For example, Woolworths caters for the high-end market, which is knowledgeable and concerned about sustainability issues, and this usually induces the company to be responsive to climate change throughout its supply chain. Woolworths has also played an important role as an 'inspector' over its supply chain to ensure that they comply with environmental standards (Heritier et al., 2009).

There is a strong representation of multinationals in this cluster of visionaries in South Africa, possibly because laws and policies such as the EU Emission Trading scheme in which these multinationals have their operations require them to comply with these regulations irrespective of where they operate and invest (Borzel and Risse, 2010). For example, SABMiller adopts a holistic approach to energy and carbon management, which applies to all their global operations. This holistic approach is embedded in their climate change target (that is, to be 50 per cent carbon efficient by 2020 over a 2008 base) which is strongly linked to phase 3 of the EU Emission Trading Scheme (SABMiller annual report, 2012). Multinationals such as Unilever and SABMiller in this cluster are also continuously targeted by NGO campaigns because of their strong brand names, which they have to defend (Honke et al., 2008). As a result, they are always proactive in responding to climate change because of concerns about their reputation.

Conclusions

This chapter outlined the mitigation strategies of the top 100 companies listed on the Johannesburg Stock Exchange in South Africa and all the listed companies on the Nairobi Stock Exchange in Kenya. The cluster analysis results of content analysis data of these companies' annual reports revealed four main clusters in South Africa (visionaries, efficiency drivers, emergent planners and laggards) and three main clusters in Kenya (efficiency drivers, emergent planners and laggards) which describe their main mitigation strategies. Due to the different operational conditions for business and varying climate change governance frameworks in many African countries it is difficult to generalise these results.

The analysis of these results reveals that performance with regard to climate change is very variable, with most companies, particularly in Kenya being low performers (that is, the laggards). The responses within the clusters are horizontal indicating that few companies perform in different dimensions. The 'visionaries', for example, consistently performed well on all the measured indicators whilst the 'laggards' consistently performed poorly on all indicators. The analysis also reveals in-country variability with regard to the responses due to sector specific institutions and corresponding perceived issue salience together with different company specific organisational resources and culture. In addition, there is also variability in responses between the countries with South African companies being more responsive than Kenyan companies because of the varying levels of limited statehood.

References

Berry, M.A. and Rondinelli, D.A. (1998) 'Proactive Corporate Environmental Management: A New Industrial Revolution', *Academy of Management Executive* 12, 2, 1–13.

Börzel, T.A., Heritier, A., Kranz, N. and Thauer, C. (2011) 'Racing to the Top? Firm's Regulatory Competition in Areas of Limited Statehood', in T. Risse and U. Lehmkuhl (eds) *Governing without a State? Governance in Areas of Limited Statehood* (New York: Columbia University Press).

Börzel, T.A. and Risse, T. (2010) 'Governance without a State: Can it Work?', *Regulation and Governance* 4, 113–134.

Carbon Disclosure Project (South Africa) (CDP SA) (2010) Carbon Disclosure Project Report 2008: JSE Top 100. Report written by National Business Initiative (NBI) and Incite Sustainability for the Carbon Disclosure Project. South Africa.

Greenwood, R., Suddaby, R. and Hinnings, C.R. (2002) 'Theorizing Change: The Role of Professional Associations in the Transformation of Institutionalized Fields', *Academy of Management Journal* 45, 1, 58–80.

Hamann, R. (2006) 'Can Business Make Decisive Contributions to Development? Towards a Research Agenda on Corporate Citizenship and Beyond', *Development Southern Africa* 23, 2, 176–195.

Hamann, R., Sinha, P., Kapfudzaruwa, F. and Schild, C. (2009) 'Business and Human Rights in South Africa: An Analysis of Antecedents of Human Rights Due Diligence', *Journal of Business Ethics* 87, 2, 453–473.

Hayes, A.F. and Krippendorff, K. (2007) 'Answering the Call for a Standard Reliability Measure for Coding Data', *Communications and Measures* 1, 1, 77–89.

Héritier, A., Mueller-Debus, A.K. and Thauer, C.R. (2009) 'The Firm As an Inspector. Private Ordering and Political Rules', *Business and Politics* 11, 4, Art.2.

Hoffman, A.J. (2010) 'Climate Change As Cultural and Behavioral Issue. Addressing Barriers and Implementing Solutions', *Organizational Dynamics* 39, 4, 295–305.

Hönke, J., Kranz, N., Borzel, T. and Heritier, A. (2008) 'Fostering Environmental Regulation. Corporate Social Responsibility in Countries with Weak Regulatory Capacities. The Case of South Africa', SFB Working Paper 9.

Kabubo-Mariara, J. and Karanja, F.K. (2007) 'The Economic Impact of Climate Change on Kenyan Crop Agriculture: A Ricardian Approach', *Global and Planetry Change* 54, 3–4, 319–330.

Ketchen, D.J., Thomas, J.B. and Snow, C.C. (1993) 'Organizational Configurations and Performance: A Comparison of Theoretical Approaches', *The Academy of Management Journal* 36, 6, 1278–1313.

Kolk, A. and Pinkse, J. (2009) 'The Influence of Climate Change Regulation on Corporate Responses: The Cases of Emission Trading', in Sullivan (ed.) *Corporate Responses to Climate Change: Achieving Emission Reductions through Regulation, Self-Regulation and Economic Incentives* (Sheffield: Greenleaf), pp. 43–47.

Kolk, A. and Pinkse, J. (2005) 'Business Responses to Climate Change: Identifying Emergent Strategies', *California Management Review* 47, 3, 6–20.

Kolk, A. and Pinkse, J. (2004) 'Market Strategies for Climate Strategies', *European Management Journal* 22, 3, 304–314.

Krippendorff, K. (1980) *Content Analysis: An Introduction to Its Methodology* (Beverly Hills, CA: Sage).

Krippendorff, K. (2004) *Content Analysis: An Introduction to Its Methodology* (2nd ed.) (Beverly Hills, CA: Sage).

Linde Group (2010) Linde Group Annual Report. Munich.

Lombard, M., Snyder-Duch, J. and Bracken, C. (2002) 'Content Analysis in Mass Communication: Assessment and Reporting of Intercoder Reliability', *Human Communication Research* 28, 4, 587–604.

Marquis, C., Zhang, J. and Zhou, Y. (2012) 'Regulatory Uncertainty and Corporate Responses to Environmental Protection in China', *California Management Review* 54, 1, 39–63.

Miles, R.E. and Snow, C.C. (1978) *Organizational Strategy, Structure and Process* (New York: Palgrave).

Miller, D. (1987) 'The Genesis of Configuration', *Academy of Management Review* 12, 4, 686–701.

Mintzberg, H. (1983) *Power in and Around Organizations* (New Jersey: Prentice Hall).

Ozawa-Meida, L., Fransen, T. and Jimenez-Ambriz, R.M. (2009) 'The Mexico Greenhouse Gas Program: Corporate Responses to Climate Change Initiatives in a Non-Annex I Country', in R. Sullivan (ed.) *Corporate Responses to Climate Change: Achieving Emissions Reductions Through Regulation; Self Regulation and Economic Incentives* (Sheffield, UK: Greenleaf), pp. 117–138.

Pervaiz, K.A. (1996) 'Culture and Climate of Innovation', *European Journal of Innovation Management* 1, 1, 30–43.

Risse, T. and Lehmkuhl, U. (2006) Governance in areas of Limited Statehood: New Modes of Governance. Research Program of the Research Centre, Working Paper No.1. SFB(700).

SABMiller Limited (2012) SABMiller Annual Report. Johannesburg.

United Nations (UN) (2009) Africa Review report on Sustainable Consumption and Production.A report by the Economic Commission for Africa, Committee on Food Security and Sustainable Development 6th session, Regional Implementation Meeting for CSD-18. Addis Abba. Available online at: http://www.un.org/esa/dsd/csd/csd_pdfs/csd-18/rims/AfricanReviewReport-on-SustainableConsumptionSummary.pdf [accessed 15 May 2010].

United Nations Development Program (UNDP) (2009) Human Development Report 2009. Overcoming Barriers: Human Mobility and development.

Available online at: http://hdr.undp.org/en/media/HDR_2009_EN_Complete. pdf [accessed 20 March 2010].

United Nations Framework Convention on Climate Change (UNFCC) 2007) Climate Change: Impacts, Vulnerabilities and Adaptation. Bonn: UNFCC.

Visser, W. (2005) 'Corporate Citizenship in South Africa: A Review of Progress since Democracy', *Journal of Corporate Citizenship* 18, 29–38.

Vogel, C. (2009) 'Business and Climate Change: Initial Explorations in South Africa', Climate and Development 1, 1, 82–97.

3
Climate Change Policies in the Car Industry: Asset Specificity as a Driver of Internal Innovation

Christian R. Thauer

Introduction[1]

This chapter investigates climate change mitigation activities in the car industry. It concentrates on internal innovations in production processes for enhanced water and energy efficiency, and on increased efficiency in other aspects of production relevant to climate change mitigation (see Figure 1.5 in Chapter 1). What drives firms' voluntary commitment to such innovations? In political science in particular, much of the literature has focused on external drivers to explain environmental firm policies and other forms of corporate social responsibility.[2] This chapter offers a complementary perspective in that it makes the case for intra-organisational factors and dynamics as drivers for contributions by business to climate change mitigation.[3] More specifically, it points out and analyses different levels of intensity of climate change mitigation among car firms in South Africa. I argue that 'asset specificity' (Williamson, 1975) is the intra-organisational feature driving climate change policies in the industry. In this context, asset specificity refers to substantial investments in production facilities and technology that are expected to pay off only in the long term (Thauer, 2010, 2013a and forthcoming 2014). While asset-specific investments may lay the foundation for a successful market strategy, they bring management into a precarious situation intra-organisationally. Asset specific investments are risky, and they are associated with uncertainty and intra-organisational information asymmetries. They make head office management dependent on the production unit that has received the investment and vulnerable to its behaviour. I show that firm policies for climate change mitigation help management to mitigate the risks, uncertainties and complexity inherent in such investments, and that they are thus a means for management to stay intra-organisationally in control, and to maintain its power position vis-à-vis the production unit that has received the investment. It is for these intra-organisational concerns with power and authority that management, in situations of asset-specific investments, insists on the adoption of climate change mitigation policies.

In what follows, I will first introduce the puzzle of differential climate change mitigation policies in the car industry, and then lay out the argument for intra-organisational dynamics and asset specificity needed to solve the puzzle. I conclude with a summary of the findings and some remarks about policy implications.

The puzzle: Climate change policies in the South African car industry

While the old triad of Europe, the US and Japan is facing a period of economic stagnation, emerging markets – such as the BRICS (Brazil, Russia, India, China and South Africa) states (Armijo, 2007) – have become the main engine of growth for the international car industry (Thauer, 2013b).[4] It is in the context of this development that the automotive industry has invested substantially in South Africa over the past decade and a half. This investment has granted the industry access to the South African growth market as well as to the US, under the African Growth and Opportunity Act (AGOA).[5] Producing in South Africa has additional advantages: labour costs are lower than in Europe, Japan or the US, and firms are less restrained by environmental regulation. South Africa is, in the field of the environment, an area of limited statehood (Hönke, Kranz et al., 2008; Lund-Thomson, 2005; Bond, 2009); environmental legislation exists on an abstract level but is often not enforced. However, the costs that labour and environmental standards impose on business do not exclusively determine the investment decisions of big car companies (Thauer, 2013b). Their overall market strategy is long-term orientated, aiming at an increase in market share. Success and failure in the industry are highly dependent on the realisation of economies of scale and technological innovation (Black, 2001; Black and Mitchell, 2002; Barnes and Black, 2003; Lorentzen and Barnes, 2004; Meyn, 2004). This makes car companies less prone to 'race to the bottom' (Chan, 2003; Singh and Zammit, 2004; Bohle, 2008) dynamics when compared to, say, firms in the textile industry, which thrive on low labour costs and a general externalisation of the costs of production (Börzel and Thauer, 2013). This chapter presents evidence showing that many of the industry's members make substantial attempts to adopt intra-firm policies that contribute to climate change mitigation. However, they do so to different degrees and at distinct points in time.

The climate change mitigation policies analyzed here in this respect are essentially environmental standards regulating the process of production ('process standards'[6]) as, for example, the management standard ISO 14001.[7] The analysis will not take into consideration 'product standards' such as, for instance, the emissions levels of cars – which are certainly the main impact of the industry with respect to climate change. Unlike process standards, product standards are often used as hidden import barriers. Therefore, government regulation and the 'shadow of hierarchy' (Héritier and Lehmkuhl, 2008; Halfteck, 2008) are here the key determinants of standard adoption (Vogel and Kagan, 2004; Vogel, 2005; Börzel et al., 2011). In contrast, what interests me in this chapter is to specify under which conditions firms pursue climate change policies in the absence of the state's regulatory influence. It is for this reason that this chapter concentrates on process standards only. In particular, I focus on standards that facilitate either an efficient use of climate change relevant resources, such as water or energy, or a reduction of emissions and waste in the process of production. Water and energy consumption, and waste and emissions outputs, do not only account for large proportions of the CO_2 footprint of automotive production. They are also cost factors in the production process that are influenced by climate change itself. Water and energy prices, for example, are likely to rise with global warming. Local governments, such as municipalities, may want to reduce waste and emissions outputs to manage their own greenhouse gas (GHG) footprint in the future and, to this end, tighten regulation in these areas, which would also lead to an increase in production costs. In these respects, the standards considered here represent not only car companies' contributions to climate change mitigation, but also the industry's attempt to deal with foreseeable political-ecological changes in the environment in which it operates.

Concerning the cases this chapter analyses,[8] the South African automotive industry is dominated by the same seven large multinational firms that dominate the global car market: BMW, Ford, General Motors, Mercedes, Nissan, Toyota, and VW. The South African branches of these seven firms are generally very similar to one another (Thauer, 2010). For example, all are instances of foreign direct investment (FDI), which makes South Africa their 'host' country, while they originate from 'home' countries (the US, Germany and Japan respectively) that tightly regulate the industry – also with respect to environmental standards. The seven firms also resemble each other in terms of size, as each of them generates revenues worth US$50–200 billion per year. In terms of budget, these companies are larger than some African states. The seven branches also have in common the fact that they operate production sites that are technologically highly advanced and complex in the context of South Africa, and that they follow a mainly export-driven business model: most of the production output is exported. While very similar in many respects, the firms show some remarkable differences concerning climate change policies.

Some firms run operations in their 'host' country, South Africa, on the basis of the same high environmental standards as are in operation at 'home'. Firm A,[9] for example, a German high-end car manufacturer, has fully implemented German environmental law regarding water usage, energy consumption, waste management and emissions in South Africa. This is remarkable as the German standards are much stricter than are those of South African law. Firm B, of South Africa, another German high-end car producer, is another case in point. This firm has also largely implemented German environmental standards in its South African production site. However, while A transferred the high standards from 'home' to its operations 'abroad' as early as the late 1990s, B did so only recently. The firm upgraded the plant in East London in 2006/07. Prior to this, production in South Africa was disproportionally energy and water intensive when compared with a German production site. Furthermore, there was no comprehensive waste management system, and the firm did not control emissions and effluents released into the environment. Hence, while the firm has environmental policies today, it did not do so to the same extent before. What explains B's time lag in this respect? Why did A feel inclined to operate on the highest attainable standards in South Africa already in the 1990s, while B was still making use of the rather lax environmental laws in South Africa?

While A and B have now implemented very high environmental standards, other firms, such as E – a US car manufacturer – continue today to operate on somewhat low environmental standards in South Africa. A production manager at E, for example, says that it takes much more energy to make a car in the Sylverton plant in Gauteng than in a comparable plant in the US or Europe. In addition, E's plant in South Africa does not manage its waste output effectively, does not account for its emissions and makes excessive use of water during the production process. Only recently has E started to improve the practices in its plant in these respects. Why is it that some firms such as A and B produce in a resource-efficient way and based on high environmental standards, while E and other firms make cars in a resource-intensive way?

To compare these differences systematically, the analysis takes into account five firms producing in South Africa: the German manufacturers of high-end cars A and B, US American mass producer E, C – a German mass producer – and D, a Japanese producer of mass market cars.[10] The analysis compares these firms with respect to the degree to which they have implemented environmental standards in the areas of water and energy consumption, as well as emission and waste output. The comparative analysis includes a cross-sectional and a longitudinal perspective. In the longitudinal perspective, the analysis subdivides the firm cases according to points in time at which the level of implemented standards and practices change. I focus, on the one hand, on procedural standards such as management systems – an example is ISO 14001 – and, on the other hand, on substantive standards that specify reduction targets or the maximum of resources that can be used over a certain period of time. Figure 3.1 illustrates the climate change mitigation policies in the South African automotive industry in comparative perspective across firms and over time. The figure details the results of the assessment of these policies, which form the dependent variable of this study. A selection of cases on this dependent variable results in

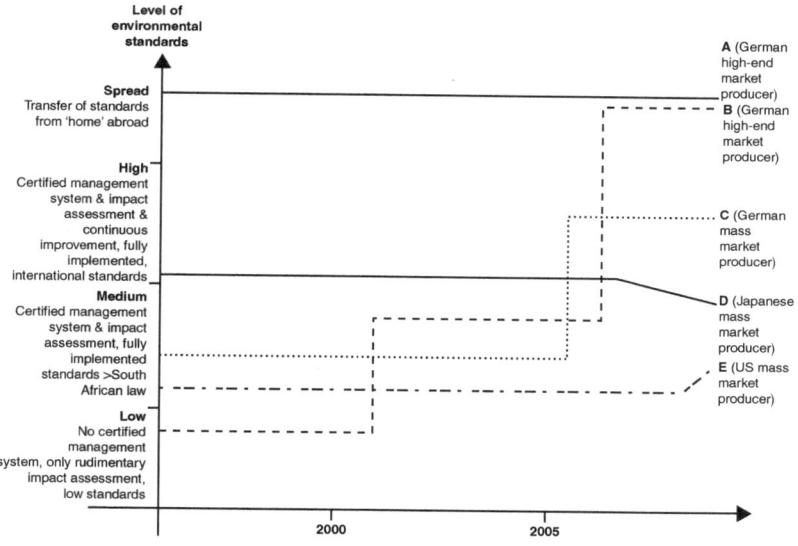

Figure 3.1   Climate change policies in the car industry (across firms and over time)

five cross-sectional cases, which make up for ten cases if we include the longitudinal perspective.

What accounts for these striking differences and trends in environmental policies across cases and over time? Chapter 1 suggests a number of factors and drivers that can push firms to adopt high standards. Can any of these arguments explain the different environmental policies among South African carmakers?

An important factor for firms' corporate social responsibility is the institutional context (Börzel, Hönke and Thauer, 2012). And an important feature of this institutional context is the regulatory threat by state agencies (that is, the 'shadow of hierarchy'), as pointed out by the literature on governance in areas of consolidated statehood (Halfteck, 2008; Héritier and Lehmkuhl, 2008). According to this literature business, when given the option, chooses not to be regulated at all. However, if this option is not available, such as when the state makes credible attempts to regulate the industry, business prefers voluntary self-regulation to state regulation. Business may therefore adopt high standards voluntarily, in order to pre-empt upcoming regulation. However, this explanation cannot account for the variation illustrated in Figure 3.1. In the policy field of the environment, South Africa has had fairly well developed legal standards on an abstract, national level since the mid-1990s. However, details pertaining to the specific behaviour of firms are often not specified on a local level (Lind-Thomson, 2005; Hönke et al., 2008; Börzel and Thauer, 2013). Overlapping responsibilities of several government departments lead to regulatory confusion, contradictory requirements and implementation gaps. Most importantly, the implementation of regulations is in many cases deficient, since local state agencies lack the capacity to effectively monitor and sanction corporate malpractice. Hence, the state is not capable of making a credible regulatory threat.

While the shadow of hierarchy is thus excluded as a potential explanation, it may be precisely the absence of state capacities in the institutional environment that motivates firms to engage in self-regulation based on a logic referred to as 'shadow of anarchy' (Börzel and Risse, 2010). Where the state is not capable of making and enforcing collectively binding decisions, no common goods may be produced at all. Some authors argue that if the pursuit of their individual profit depends on the provision of certain common goods and collectively binding rules to produce them, and the state is incapable or unwilling to provide these, companies have a major incentive to fill the governance gap (Börzel and Risse, 2010). However, the business model of the featured firms is highly similar across the ten cases. The motivation to step forward as governance provider is thus the same for all cases, and thus cannot explain the different activities of the firms in this respect. Also, environmental standards regulating the process of production usually do not create a common good which would be otherwise lacking, but rather define non-excessive appropriation of public goods (such as clean air, clean rivers, electricity supply etc.). In other words, firms' ability to produce is not impinged upon by the lack of environmental regulation and its enforcement (or it is affected only very indirectly). Therefore, this argument cannot account for the variation across the ten cases with respect to the featured firms' level of environmental standards.

Another important argument for climate change policies pertains to reputational concerns (Börzel and Thauer, 2013). Campaigns and disclosures organised by NGOs are a factor in this respect (Spar and LaMure, 2003; Hendry, 2006; Schepers, 2006). Generally, the five large multinational car firms featured in the analysis above are under pressure from NGOs, but only for the cars they produce or, to be more precise, for the emissions levels of those cars. They are not usually targeted for their production processes. This is confirmed not only by managers of the five firms, who say they do not feel under pressure from NGOs regarding production processes,[11] but also by a NGO representative, who says, 'regarding their production processes, I fully trust them. In this respect, they are way above anything in South Africa. So we do not target them in this area. We only have problems with their products'.[12] Hence it is not NGO pressure that motivates car companies to engage in environmental policies; this factor does not explain the observed variation between the ten firm cases with respect to their efficiency in resource use.

Another reputational concern that may motivate climate change policies derives from the market orientation of firms and corresponding consumer expectations and pressure (Spar and LaMure, 2003; Smith, 2008; Héritier, Müeller-Debus and Thauer, 2009). Whereas the German producers A and B are high-end market firms, Japanese D, German C and US American E target the mass market. However, this difference in market orientation also cannot explain the asserted variation. The high-end market firm B operated for a long time on environmental standards at a level similar to the ones of the mass market producers C and E – and even below the mass market firm D's. Differences in the target market of the firms can also not explain why the high-end market firm B began, in 2006/07, to upgrade its operations with respect to environmental standards, or why the mass market manufacturer D's environmental policy has recently weakened. Hence, this factor also does not account for the asserted variance between the ten cases of different levels of environmental policies in the automotive industry.

Other potential drivers of climate change mitigation are various forms of 'normative institutions', such as between firms or between firms and government or NGOs. For example, business associations can facilitate self-regulation of firms (Ronit and Schneider, 2000), 'green clubs' (Prakash and Potoski, 2006) such as ISO 14001 can provide firms with incentives to adopt environmental standards, and so can public–private partnerships. However, these cannot account for the asserted variation either. The ten cases shown in Figure 3.1 strongly resemble one another with respect to the firms' membership and participation in institutions of cooperation. They are members of identical associations in South Africa, are exposed to the same extent to the emergence of ISO 14001 internationally, and usually do not cooperate with state agencies in the context of public–private partnership, with respect to their processes of production. Therefore, this explanation also cannot account for the finding of differential climate change policies in the South African car industry. What, then, is the explanation? I shall argue in the following that intra-organisational asset-specific investments motivate firms to adopt climate change policies.

Asset specificity as a driver of climate change policies

Asset specificity refers to non-redeployable investments (Williamson, 1975, 1996). However, non-transferability is rarely absolute. Another way to describe asset specificity is therefore that the investments made to support a particular transaction have a higher value in relation to that transaction than they would if they were redeployed for another purpose (McGuinness, 1994). Originally developed to explain the 'make or buy' decision in market exchanges, the concept is applied in this chapter intra-organisationally, to explain why and under which conditions firms adopt climate change policies. More specifically, I argue that substantial investments in the production unit, and with a long duration time before they are envisioned to pay off, have such effects (see for the following Thauer, 2010, 2013a and forthcoming 2014). Asset specificity has therefore two dimensions. The asset dimension refers in this context to the relative[13] quantity of resources delegated to the production unit to set up and run production. The more resources are allocated, the higher the potential level of asset specificity. The specificity dimension entails a temporal commitment, that is, the period before envisioned return of investment, because during that period management cannot redirect resources to support another task or to build up production capacities elsewhere. The longer the commitment of resources, the higher is the degree of specificity of – and dependence of the firm on – the given asset. The absence of high values in either of the two dimensions – i.e. of significant investments or a commitment – indicates the absence of asset-specific allocation of resources. If only few resources are allocated to support the start up of a new production facility, management is not dependent on the production unit. Wherever investments are made, if there is little commitment they can easily be redirected to support the creation of production capacities elsewhere. Therefore, resource allocation is, in these cases, not asset specific. Asset-specific allocation of resources in production, instead emerges with high values in both dimensions – the more resources are allocated to support the creation of production facilities, and the longer the period in which these resources cannot be transferred due to a long commitment, the higher the degree of asset specificity.

But why should asset specificity make firms adopt climate change policies? Intra-organisationally, asset specificity creates pockets of expert knowledge inside the production unit that has received the investment. This results in information asymmetries in the relation between managers in the head office and the production unit (Thauer, 2010, 2013a and forthcoming 2014). However, information asymmetries give rise to 'hidden action' (Miller, 1992), opportunism, shirking and cheating (Williamson, 1975), which in turn threaten the authority of management in the head office and the economic efficiency of the whole firm. In addition, asset specificity creates vulnerabilities in the sphere of firm–society relations. Once resources have been allocated, the firm becomes vulnerable to changes in the environment, as management in the head office is subsequently circumscribed in its ability to react to such changes through swift reallocation. Asset specificity thus puts these managers in the head office in a precarious position.

But why should this make them opt for climate change policies? Can such policies improve the position of managers in situations of asset-specific investments vis-à-vis the production unit that has received the investment? And if so, how and why? I argue that these policies entail standards regulating the process of production that help management, first, to stay in control and maintain its power position vis-à-vis subunits, and second, to reduce its vulnerability to the investment environment (Thauer, 2010, 2013a and forthcoming 2014). In the first argument, environmental standards reduce information asymmetries, in particular as these standards are usually implemented by management systems such as ISO 14001. Management systems structure, document and monitor work processes, identify areas of improvement, provide for performance measures and facilitate resource use efficiency. Hence, they enable management to collect information about what subordinates do, according to which standards they can be held accountable, and how processes can be improved.

In the second argument, environmental standards that contribute to climate change mitigation reduce uncertainty. Firms that invest in an asset-specific way worry about whether their investment decision will continue to be valid in the future. They ask themselves, for instance, whether energy, waste or water prices will increase, or whether the firm will be forced to reduce emissions (for example, through future NGO campaigns, which cannot be excluded, or increasing prices due to a shortage of resources or the potential of future taxes). However, firms that adhere to high standards of climate change mitigation from the very start reduce this kind of vulnerability to changes in the environment – they have prepared for such changes already. Therefore, a management that has an asset-specific relationship with production, and thus cannot easily close down a unit to shift its strategic focus, will insist on implementing strict standards of climate change mitigation to reduce its vulnerability to the local investment environment.

I shall now turn to the empirical analysis of these arguments in order to solve the puzzle. It begins with a consideration of the case of B, a large German automotive company, which is also a global, high-end market brand. B makes up for three cases if we include the longitudinal perspective. More precisely: case 1 is B at t1 during the late 1990s (1996–2001/02), case 2 is the firm at t2 between 2002 and 2006/07, and case 3 at t3 in the years 2006/2007 and onwards.

B at t1 illustrates no particular degree of asset specificity as concerns the investments made in the South African branch. During the 1990s the firm produced mainly in 'CKD-mode' in South Africa.[14] 'CKD' refers to 'completely knocked down' vehicles. 'CKD-mode' denotes a process according to which new cars made in Germany are disassembled and shipped to South Africa only to be reassembled in B's East London plant for sale in South Afica.[15] The assembly line required ongoing investment of up to 10 million US dollars per year, which equals about 0.5 per cent of the total annual investment volume of B globally.[16] In comparative perspective, these investments are smaller than those at t2 and t3 (see below). Most important, however, the length of time before these investments were expected to pay off was fairly short at that time. They led to immediate increases in assembly capacity, which directly translated to bigger sales capacities. Hence, they generated higher returns at once. These investments were therefore not very asset specific and, accordingly, they did not trigger any significant climate change policies – at least not when compared with B at t3 or A. For example, during t1 the firm did not have a third-party audited environmental management system such as ISO 14001 in place; it did also not measure its impact in the areas of water and energy consumption; and it did not measure its emissions output or engage in waste reduction or recycling.

Over time, however, the South African branch emerged as an increasingly important export hub for the mother company's headquarters. In response to this development, the operation in South Africa was developed in two major steps from a mere assembly plant into a full production site with significant production depth. For these upgrades, the German headquarters had to make significant investments in its South African production branch with long periods before returns were expected. The level of asset specificity increased. More specifically, at t2 the decision was made to upgrade B's South Africa assembly plant to a plant with production components. To this end, investments of about 30 million US dollars were made on average each year,[17] equalling 1 to 1.5 per cent of the firm's global annual investment volume.[18] Returns on these investments were not expected for about two to three years. Hence, when compared to t1, the case of B at t2 is marked by higher levels of asset specificity. The climate change policies of the firm at t2 are concomitantly stronger. Immediately before the first asset specific investments were made, in 2001/02, the firm decided to enter the ISO 14001 certification process, and this was achieved in 2003. The management system was, according to representatives of B, not particularly well implemented during that time. Nonetheless, ISO 14001 certification means that, unlike at t1, at t2 B had a third-party audited environmental management system in place. During t2 B also began to assess its impact in terms of water usage and energy consumption, which it had not done during t1. While the firm did not (yet) implement strict reduction targets or transfer water and energy consumption standards from Germany to its South African operation, it made sure that its business practices in these areas were in accordance with the South African law – which is, however, fairly lax and shallow when compared to the law in, for example, Europe.

At t3, then, the firm's stance towards climate change policies dramatically changed yet again, due to another drastic increase in asset specific investments. B of South Africa won an internal tender for the production of the next generation of the entry-level model of the company.[19] The new contract was settled with central management in Germany and triggered unprecedented investments for the South African branch. A new factory was built specifically for the production of the new generation model, with 250 robots (the old plant had only six robots) and a renewed paint shop (the ovens that dry and harden the paint were technically upgraded). The branch was transformed from an assembly plant with some manufacturing components (t2) into an original equipment manufacturer-site marked by high degrees of asset-specific investments (t3). To this end, headquarters invested about US$300–350 million in 2006/07 in the branch.[20] Nine to 11 per cent of the annual total global investment volume of B went into South Africa in both years. The construction of the new plant took one and a half years. A production manager at B estimated that it would take another two to two and a half years of production in the new plant before return of investment. This increase in asset specificity – in particular the long period (3–4 years) before envisioned return of investment – inspired management in Germany to insist on the implementation of strict climate change policies. Even before the first investments were made, headquarters sent more than 20 production and environmental engineers from Germany to South Africa, with the task of implementing German standards for water and energy consumption, waste and affluent management, and emissions. As a consequence, the branch has operated since 2006 using the environmental regulations of the municipality of Stuttgart. This is remarkable, particularly since the South African operations are located in East London in the Eastern Cape province, which has lax regulation and weak enforcement capacities even by South African standards. In addition, the team of environmental engineers who were sent to South Africa made the branch install the ISO 14001 management system again and according to stricter standards, in order to achieve higher levels of implementation, in particular the continuous improvement processes that are an integral part of the management system. The firm also implemented at that time the standard VDA 6.1[21] – an integrated environmental and quality management system on a level above ISO 14001. Hence, in terms of procedural standards (that is, ISO 14001 and VDA 6.1) and substantive standards (Stuttgart's water, energy consumption, emissions and waste management regulations), since t3 the firm operates not only on a higher level than at t1 and t2, but also on a level virtually unmatched by any South African company practices.

Asset specificity appears to explain this strengthening of climate change policies over time. The production and environmental manager of B confirmed this indirectly. When asked why B voluntarily imple-ments German environmental regulations in its East London plant, he responded that for management in Germany this is a way of dealing with the tremendous risks inherent in the massive investments made in the South African plant from 2006 on. To assure that 'the house is in order' and that resources are used efficiently, headquarters insists on the implementation of high levels of environmental policies before it is ready to actually invest in South Africa.[22] These comments support the theoretical link between asset specificity and climate change policies, and the use of the latter as a means of risk mitigation and a means of maintaining control over the use of scarce organisational resources.

The case of the German car firm A, which is also a global high-end brand, is worth considering in comparison with the three cases of B (t1–t3). The case – case 4 in the analysis so far – illustrates asset specificity on a constant high level (see Figure 3.1). A had already decided in the mid/late 1990s to upgrade its branch in South Africa to a full production plant, and to this end the firm made investments that were similar to B in t3 in terms of both the size of the investment and the time before expected return. In 1998, for example, A invested about US$200 million in South Africa (equalling approximately 10 per cent of the firm's total investment volume at the time) with a duration of about three to four years before returns were expected to outgrow the initial investment.[23] Similarly high investments were made again in 2004/05, again with three to four years before returns were expected to have paid off the investment. Hence, from the mid/late 1990s onwards, the case of A is marked by similarly high levels of asset specificity as the case of B at t3. As in the case of B, the high levels of asset specificity are reflected in the firm's climate change policies: A has had VDA 6.1 and a strict ISO 14001 implementation, as well as aggressive energy and water reduction targets, from the mid/late 1990s. In fact, the firm had already implemented the environmental regulation of the municipality of Munich in South Africa. This is not only true for water and energy consumption, but also for waste management and emissions limitations. In addition to providing additional evidence for the link between asset specificity and climate change policies, the case can also be used as a control case for the lon-gitudinal analysis of B. The greater engagement in environmental poli-cies of B between t1, t2 and t3 could be the result of a general historical 'trend' in the industry, as 'diffusion theory' (Prakash and Potoski, 2006) would suggest – and turn out not to be the outcome of higher degrees of asset specificity. However, since A's environmental policy is constant throughout these periods, this alternative explanation can be dismissed.

Looking next at German C and American E, these two mass market producers make up four cases. The German multinational C produced mainly in CKD mode until 2006, just as E did until 2009. The two cases (C until 2006 and E until 2009) are cases five and six in this analysis. CKD-mode production required constant investments from both firms, but on a fairly low level and with short periods before returns were generated to an extent so that they would pay the investment off. As mentioned above, investments in assembly plants usually directly translate into higher and more efficient production output – and, hence, higher sales volumes.[24] Hence, the level of asset specificity that characterises the two cases is rather limited, and so is their level of climate change policies. While both C and E, obtained ISO 14001 certification in the years 1998/99, the management system is rather weakly implemented and based on lax standards. At the plant of E, for example, 'littering and waste production is, as you have seen [on a factory plant tour], quite a serious problem on our site'.[25] Also, E uses 600 per cent more energy manufacturing a car than would a comparable plant in Europe, Asia or the US and also makes excessive use of water. The findings for C are similar in this respect. Both firms, however, also did not control their emissions output during that time. Hence, E until 2009 and C until 2006 operate on rather lax environmental standards in the climate change relevant areas of energy and water consumption as well as waste management and emissions outputs.

At C this started to change in 2006, after the South African branch won a new bid from C Germany. C since 2006 is thus the seventh case considered in the analysis. The South African branch was subsequently upgraded through major investment. For example, a new paint shop and an entirely new production line were installed in 2007. In the context of these investments, the plant was developed from a purely CKD-mode assembly plant to a factory with a production mix of 'CKD-mode' assembly and original equipment manufacturing.[26] Since then C is accordingly marked by significantly higher levels of asset specificity. The new paint shop, for example, cost about 10 million US dollars, and it is part of a US$60–70 million investment package made by C Germany in 2006/07 to technologically upgrade and enlarge the South African branch. The average time before return of investment of this US$60–70 million is about two years. A similar development started at E in 2008/09, which is the eighth case of the analysis. The branch won a bid from E America, which requires the headquarters in Detroit to make asset-specific investments in South Africa. More specifically, E had announced annual investments of up to US$70 million until

2010/11 in new technology, a new paint shop and installation of a new production line producing utility vehicles from 2010 or 2011 onwards. These investments had a time to return of investment similar to the investments of C from 2006 onwards and thus raised the level of asset specificity of E significantly.

Since the announcement, the environmental policies of E changed drastically. The environmental and production managers re-implemented the ISO 14001 management system in order to gain control over the plant's littering problem and to implement basic principles of re-use and recycling. In addition, global headquarters forced the branch to reduce, in two consecutive years, the yearly consumption of energy and water by 17 per cent. Otherwise, headquarters refused to invest in South Africa. When asked why headquarters would insist on these reduction targets, the manager said that for such long-term and substantial investments as the ones envisioned over the next few years, resource use efficiency and tightly regulated production processes – as facilitated by a strict implementation of ISO 14001 – are a precondition.[27] Otherwise, such investments would appear too risky – a statement confirming what has been theorised here – environmental standards help management to deal with the uncertainties and complexity of asset specific investments.

To summarise, the cases of C and E are additional evidence for the argument that asset specificity triggers climate change policies. Before 2006 and 2008 respectively, the two firms did not receive significant asset specific investments – and their environmental standards in the climate change relevant areas of water and energy consumption as well as emission outputs and waste management were accordingly shallow. When the level of asset specificity was raised because of substantial investments with a long duration, the South African branches of the firms were forced by their respective headquarters to strengthen their environmental policies in these areas as well. This relation between asset specific investments and climate change policies is also illustrated by the case of the Japanese mass market producer D – a particularly interesting case, as the factory's environmental policy was in decline between 2005/06 to 2008/09.

Before 2006, D planned to upgrade the South African plant from an assembly plant into an assembly plant with a production component – not unlike C in 2006 or E. To this end, the branch was expected to receive asset specific investments such as a paint shop and a new production line. Accordingly, the Japanese headquarters started to demand strict ISO 14001 environmental management implementation and began to formulate aggressive reduction targets in the areas of energy and water consumption.[28] However, in the end, the envisioned investments were never made; the CEO of the firm since 2001 decided to cut the investments as part of his plan to make D profitable by 2005. D had to let go employees in Japan to meet this target, which caused a public outcry. In the context of these austerity measures, investments in South could not be justified and were therefore scrapped altogether in 2005/06. The consequence of these developments is that the environmental management system ISO 14001 has been in decline since then, in terms of the strictness of its implementation, though the factory in South Africa is still certified. Also, the reduction targets of water and energy use and emissions output as well as recycling were not set by the Japanese headquarters any more – and therefore declined to the level demanded by South African law. The two cases – case nine (before 2005/06) and ten (after 2005/06) in the analysis – of D South Africa thus illustrate that the relation between asset specificity and environmental standards is valid in both ways. The first eight cases analysed above show that increases in asset specificity lead to stricter environmental standards. The last two cases of D over time demonstrate that declining levels of asset specificity lead to a parallel decline in environmental policies. I therefore conclude that asset specificity solves the puzzle of differential climate change mitigation policies in the South African automotive industry.

Conclusion

This chapter analysed voluntary climate change mitigation measures in the car industry of South Africa. The empirical inquiry showed that factors in the environment of car firms, such as NGO pressure, the level of regulation in their country of origin, the 'shadows' of hierarchy and of anarchy, consumer orientation, membership in norm-promoting initiatives, cannot account for this variation. Instead, intra-organisational asset-specific investments in production units explain why, when and how firms take climate change mitigation measures. Asset specificity refers in this context to substantial investments in production facilities and technology with a long payoff period. Asset specificity increases the uncertainty of managers in head offices with respect to the investment environment, and information asymmetries in relation to the production unit. Climate change policies help management mitigate these risks. It is for these concerns of managers in head offices with economic efficiency and intra-organisational power politics that climate change mitigation measures emerge as a consequence of asset specific investments.

These findings are also important for policy makers and managers in firms, as they point to a rationale for industry-driven climate change policies that has not been looked at much before. Environmental activists, policy makers and those driving corporate social responsibility in the business world – i.e. environmental managers and managers in associations – may use it to make sceptics among top managers in firms see the benefits of climate change mitigation.

Notes

1. I would like to thank Tanja A. Börzel and Ralph Hamann for helpful comments on an earlier draft of the chapter. Special thanks go also to the participants of a common colloquium of the Center for Transnational Relations, Foreign and Security Policy and the Center for European Integration at the Freie Universität Berlin, and in particular to Thomas Eimer, for a helpful discussion of an early draft of the paper.
2. Studies point to the 'shadow of hierarchy' (Halfteck 2008; Héritier and Lehmkuhl 2008) and, especially in 'areas of limited statehood' (Risse 2011; Börzel and Thauer 2013), where the shadow of hierarchy is absent, to the importance of pressure from consumers (Smith 2008), NGOs (Spar & LaMure 2003; Hendry 2006; Schepers 2006) and associations (Cutler et al. 1999; Ronit & Schneider 2000; Hall & Bierstecker 2002), and the reputation that standards provide (Prakash & Potoski 2006).
3. Thereby building on the works of Prakash (2000), Gunningham et al. (2003), Howard-Grenville (2007) and Thauer (2010, 2013a and forthcoming 2014); I shall also mention that in econometric analyses some internal factors are included as control variables (e.g. Khanna et al. 2007).
4. BBC News, 'Globalizing the Car Industry' (by Steve Schifferes); http://news.bbc.co.uk/2/hi/business/6346325.stm (accessed 22-Jan-2012). Growth rates of 20–25% between 2004 and 2011 make South Africa one of the fastest-growing car markets world wide (see NAAMSA 2011).
5. See www.agoa.info/index.php?view=country_info&country=za (accessed 23-Jan-2011). South African car exports grew from about 9,000 cars in 1995 to 58,000 in 2000; 114,000 cars were exported in 2005, increasing to 170,000 in 2007 and 280,000 in 2011. See NAAMSA 2006 and 2011.
6. Concerning the distinction between process and product standards see Vogel 1995: 18, and Vogel 2005.
7. ISO 14001 management systems structure work and production processes with respect to environmental aspects on the basis of an environmental impact assessment.
8. The data on which the analysis draws on consist of personal interviews with managers, firm representatives and stakeholders (such as NGOs, Union representatives, government officials, representatives of associations, experts and competitors). Additional information is provided by financial and corporate social responsibility reports of the five firms, journalistic output and

academic publications. Data were collected in the period between February 2007 and December 2008. The analysis therefore does not cover any developments in the South African car industry afterwards.

9. For confidentiality the firm names are abbreviated with a designated capital letter throughout the article.
10. Unfortunately, two firms were unwilling to participate in an interview, and information on those two companies is thus too incomplete to be included in the analysis.
11. Interview with the Manager: Quality Management of B South Africa, 23 Sept 2008; with the General Manager, Corporate Planning of A South Africa, 20 Feb 2007, Midrand.
12. Interview with the Automotive Sector Specialist of the WWF South Africa, 29 Sept 2008, Johannesburg.
13. Relative as measured in proportion to the size of the firm or its overall investment volume.
14. Interview with the Manager: Quality Management and Integrated Management System of B South Africa, 23 Sept 2008, East London; with the Corporate Health Manager of B South Africa, 26 Feb 2007, East London.
15. There are three reasons for this: first, import restrictions and local content requirements force firms to add some local content (usually of minor importance) to the vehicle. Second, there are a number of countries that grant Africa preferential access to their markets. To have an African production base allows auto firms to enter these markets on lower duties. Third, the global car sales market is restricted by import and export quotas. A CKD-site in South Africa enables a German firm, for example, to export more to Asia by using South Africa's export quotas in addition to the German ones.
16. Email questionnaire answered by the Quality and Integrated Systems Management of B South Africa.
17. Interview with the Manager: Quality Management and Integrated Management System of B South Africa, 23 Sept 2008, East London; email questionnaire answered by the Quality and Integrated Systems Management of B South Africa.
18. Annual Report of B, year 2000.
19. Interview with the Manager: Quality Management and Integrated Management System of B South Africa, 23 Sept 2008, East London.
20. Ibid.; Annual Report of B, year 2000.
21. A standard developed in the context of the German association 'Verband der Automobilindustrie', VDA, which was subsequently developed into the now widely adopted standard TS 16949.
22. Interview with the Manager: Quality Management and Integrated Management System of B South Africa, 23 Sept 2008, East London.
23. Interview with the Project Leader VPS, the SHE Representative, and the General Manager Corporate Planning of A South Africa, 20 Feb 2007, Midrand; with the Occupational Health Manager of A South Africa, 14 Feb 2007, Rosslyn.
24. Interview with the Manager for Employee Wellness and Occupational Health of E South Africa, 14 Sept 2007, Silverton; the Manager for Environmental and Safety Engineering of E South Africa, 20 Feb 2007, Silverton; the Quality and Environmental Systems Engineer of E South Africa, 30 Sept 2008,

Sylverton; the Environmental Controller for Manufacturing Planning of C South Africa, 22 Sept 2008, Uitenhage.
25. Interview with the Quality and Environmental Systems Engineer of E South Africa, 30 Sept 2008, Silverton.
26. Interview with the Health & Safety Manager, and the Environmental Controller for Manufacturing Planning of C South Africa, 22 Sept 2008, Uitenhage; interview with the Occupational Health, Employee Wellness and HIV Manager of C South Africa, 25 Sept 2007, Uitenhage.
27. See interview with the Quality and Environmental Systems Engineer of E South Africa, 30 Sept 2008, Silverton.
28. Interviews with the CSR Manager, the Manager Integrated Systems and the Senior Manager in Corporate Affairs, D South Africa, 14 Feb 2007, Rosslyn.

References

Armijo, E.L. (2007) 'Special Edition: The BRICs Countries (Brazil, Russia, India, and China) in the Global System', *Asian Perspective* 31, 4.

Barnes, J. and Black, A. (2003) Motor Industry Developing Program, *Review Report* (Durban and Cape Town).

Black, A. (2001) 'Globalization and Restructuring in the South African Automotive Industry', *Journal of International Development* 13, 779–796.

Black, A. and Mitchell, S. (2002) 'Policy in the South African Motor Industry: Goals, Incentives, and Outcomes', *The South African Journal of Economics* 70, 8, 1273–1297.

Börzel, T.A. and Thauer, C.R. (eds) (2013) *A Race to the Top? Business and Governance in South Africa* (Houndmills: Palgrave).

Börzel, T.A. and Risse, T. (2010) 'Governance without a State – Can It Work?', *Regulation and Governance* 4, 2, 1–22.

Börzel, T.A., Héritier, A., Kranz, N. and Thauer, C.R. (2011) 'Racing to the Top? Regulatory Competition among Firms in Areas of Limited Statehood', in T. Risse (ed.) *Governance without a State? Policies and Politics in Areas of Limited Statehood* (New York: Columbia University Press).

Börzel, T.A., Hönke, J. and Thauer, C.R. (2012). 'Does It Really Take the State? Limited Statehood, Multinational Corporations, and Corporate Responsibility in South Africa', under review with *Business and Politics*.

Bohle, D. (2008) 'Race to the Bottom? Transnational Companies and Reinforced Competition in the Enlarged European Union', in B. Van Apeldoorn, J. Drahokoupil and L. Horn (eds) *Neoliberal European Governance and Beyond – The Contradictions and Limits of a Political Project* (London: Palgrave), pp. 163–186.

Bond, P. (2009) 'Social Movements and Corporate Social Responsibility in South Africa', *Development and Change* 39, 6, 1037–1052.

Chan, A. (2003) 'A Race to the Bottom', *China Perspectives* 46, 41–49.

Cutler, C.A., Haufler, V. and Porter, T. (eds) (1999) *Private Authority and International Affairs* (Albany: State University of New York).

Gunningham, N., Kagan, R. and Thornton, D. (2003) *Shades of Green. Business, Regulation, and Environment* (Stanford: Stanford University Press).

Halfteck, G. (2008) Legislative Threats. *Stanford Law Review* 61 (available at SSRN: HYPERLINK. http://ssrn.com/abstract=1113173).

Hall, R.B. and Bierstecker, T.J. (eds) (2002) *The Emergence of Private Authority in Global Governance* (Cambridge: Cambridge University Press).

Hendry, J. (2006) 'Taking Aim at Business', *Business and Society* 45, 1, 47–86.

Héritier, A. and Lehmkuhl, D. (eds) (2008) 'The Shadow of Hierarchy and New Modes of Governance', *Journal of Public Policy*, special issue, 28, 1.

Héritier, A., Müller-Debus, A. and Thauer, C. (2009) 'The Firm As an Inspector: Private Ordering and Political Rules', *Business and Politics* 11, 4, 1–32.

Hönke, J., Kranz, N., Börzel, T.A. and Héritier, A. (2008) 'Fostering Environmental Regulation? Corporate Social Responsibility in Countries with Weak Regulatory Capacities. The Case of South Africa', SFB 700 Working Paper 9.

Howard-Grenville, J. (2007) *Corporate Culture and Environmental Practice. Making Change at a High-Tech Manufacturer* (Cheltenham: Edward Elgar).

Khanna, M., Koss, P., Jones, C. and Ervin, D. (2007) 'Motivations for Voluntary Environmental Management', *Policy Studies Journal* 35, 4, 751–772.

Lorentzen, J. and Barnes, J. (2004) 'Learning, Upgrading, and Innovation in the South African Automotive Industry', *The European Journal of Development Research* 16, 3, 465–498.

Lund-Thomson, P. (2005) 'Corporate Accountability in South Africa. The Role of Community Mobilizing in Environmental Governance', *International Affairs* 81, 3, 619–633.

McGuinness, T. (1994) 'Markets and Managerial Hierarchies', in G. Thompson, J. Frances, R. Levacic and J.C. Mitchell (eds) *Markets, Hierarchies and Network* (London: Sage), pp. 66–81.

Meyn, M. (2004) 'The Export Performance of the South African Automotive Industry. New Stimuli by the EU–South Africa Free Trade Agreements?', *Berichte aus dem Weltwirtschaftlichen Colloquium der Universität Bremen* 89.

Miller, G.J. (1992) *Managerial Dilemmas. The Political Economy of Hierarchy* (Cambridge, NY: Cambridge University Press).

NAAMSA (2011) Media Release: Comment of the October 2011 New Vehicle Sales Statistics: http://www.naamsa.co.za/flash/press.htm [accessed: 05 November 2011].

NAAMSA (2006) Media Release: Industry Vehicle Sales, Export and Import Data 1995–2007: http://www.naamsa.co.za/papers/20060124/export_import_1995_2007.htm [accessed 1 09 2006].

Prakash, A. (2000) *Greening the Firm. The Politics of Corporate Environmentalism* (Cambridge: Cambridge University Press).

Prakash, A. and Potoski, M. (2006) *The Voluntary Environmentalists: Green Clubs, Iso 14001 and Voluntary Environmental Regulations* (Cambridge: Cambridge University Press).

Risse, T. (ed.) (2011) *Policies and Politics in Areas of Limited Statehood* (New York: Columbia University Press).

Ronit, K. and Schneider, V. (2000) 'Private organizations and their contribution to problem solving in the global arena', in K. Ronit and V. Schneider (eds), *Private Organizations in Global Politics* (New York: Routledge), pp. 1–34.

Schepers, D.H. (2006) 'The Impact of NGO Network Conflict on the Corporate Social Responsibility Strategies of Multinational Corporations', *Business and Society* 45, 3, 282–299.

Singh, A. and Zammit, A. (2004) 'Labour Standards and the "Race to the Bottom": Rethinking Globalization and Workers' Rights from Developmental and Solidaristic Perspectives', *Oxford Review of Economic Policy* 20, 1, 85–104.

Smith, C.N. (2008) 'Consumers As Drivers of Corporate Social Responsibility', in A. Crane, A. McWilliams, D. Matten, J. Moon and D.S. Siegel (eds) *The Oxford Handbook of Corporate Social Responsibility* (Oxford: Oxford University Press), pp. 303–323.

Spar, D.L. and LaMure, L.T. (2003) 'The Power of Activism: Assessing the Impact of NGOs on Global Business', *California Management Review* 45, 78–101.

Thauer, C.R. (2010) Corporate Social Responsibility in the Regulatory Void – Does the Promise Hold? Self-Regulation by Business in South Africa and China. Phd Thesis. Ph.D. diss., European University Institute, Florence.

Thauer, C.R. (2013a). 'Goodness Comes From Within. Intra-organizational Dynamics of Corporate Social Responsibility', *Business and Society, Online First:* http://intl-bas.sagepub.com/content/early/2013/04/17/0007650313475770. full.pdf+html (24th April 2013).

Thauer, C.R. (2013b) 'Upgrading the Periphery? The Contribution of Car Companies to Environmental Governance in South Africa', in T.A. Börzel and C.R. Thauer (eds) *A Race to the Top? Business and Governance in South Africa* (Houndmills: Palgrave).

Thauer, C.R. (forthcoming 2014). *Internal Drivers of Corporate Social Responsibility. Managerial Dilemmas and the Spread of Global Standards* (Cambridge: Cambridge University Press).

Vogel, D. (1995) *Trading Up: Consumer and Environmental Regulation in a Global Economy* (Cambridge: Harvard University Press).

Vogel, D. (2005) *The Market for Virtue. The Potential and Limits of Corporate Social Responsibility* (Washington, DC: The Brookings Institution).

Vogel, D. and Kagan, R. (eds) (2004) *Dynamics of Regulatory Change: How Globalization Affects National Regulatory Policies* (Berkeley: University of California Press).

Williamson, O. (1975) *Markets and Hierarchies: Analysis and Antitrust Implications. A Study in the Economics of Internal Organization* (New York: The Free Press/ Macmillan).

Williamson, O. (1996) *The Mechanisms of Governance* (New York: Oxford University Press).

4
Renewable Energy Incentives across Varying Levels of Statehood

John Fay

Introduction

Climate change governance includes contributing towards mitigating greenhouse gas (GHG) emissions in order to maintain a stable climate as an essential public good. This chapter explores the mitigation aspect of climate change governance by assessing market-based incentives (MBIs) that promote renewable energy. This research illustrates the central role that statehood plays in renewable energy development, specifically with regard to financing considerations, which are generally considered separately from socio-political factors. The private sector independent power producer (IPP) is often dependent on the state, to provide and maintain financial incentive over the lifetime of the project, in order to ensure a sufficient return on investment for its renewable energy projects. Conversely, the state is dependent on the IPP to deliver clean energy to the country's electricity supply at the lowest possible cost, in order to maximise use of limited governmental resources. However, in developing countries, MBIs also need to provide investors with a return commensurate to the risk associated with a lower level of statehood, thus serving as the functional equivalent to the state's shadow of hierarchy (Börzel and Risse, 2010).

The objective of this chapter is to identify how financial considerations affect the deployment of renewable energy in developing countries. This research analyses the cost of capital required for the renewable energy projects to establish a quantifiable proxy of the risk profile that is determined in large part by the countries' level of statehood. In order to build the theory, case studies of South Africa and Germany were drawn from a non-random theoretical sampling (Eisenhardt and Graebner, 2007; Eisenhardt, 1989). The case studies focus on the various contextual features that influence project finance, as illustrated in hypothetical wind farm projects. The two countries were selected because both have implemented MBIs to promote renewable energy. Germany and South Africa also differ significantly, with regard to their degree of statehood and level of socio-economic development. South Africa is an emerging economy with a nascent wind energy sector, while Germany is a developed country with an advanced wind energy industry. Data has been collected through literature review, document analysis and interviews with renewable energy sector stakeholders in Germany and South Africa.

In the remainder of this chapter, the background of renewable energy MBIs and considerations for renewable energy in developing countries are discussed. Next, a comparative analysis of hypothetical wind energy projects in South Africa and Germany will be made, with an emphasis on the cost of capital resulting from variance in the level of statehood of the host country. Finally, international and national policies to adapt MBI design to the host country's level of statehood are discussed.

Renewable Energy Incentive Background

Limiting the long-term average temperature increase to 2 °C, commonly correlated with CO_2 parts per million (ppm) of 450, has been established as the consensus mitigation target to maintain a stable climate (Pachauri and Reisinger, 2007; IEA, 2008; IEA, 2011). Unfortunately, by not fully accounting for the negative externalities to the environment, carbon-intensive power generation benefits from artificially low tariff prices that in part help to create a cost advantage over renewable-based electricity generation (Menanteau et al., 2003; Dinica, 2006). As a result, it is anticipated that the global emissions trajectory will go well beyond the target 450 scenario.

The World Energy Outlook 2011 sounds an alarm with the following prediction: 'If stringent new action is not forthcoming by 2017, the energy-related infrastructure in place will generate all the CO_2 emissions allowed in the 450 Scenario up to 2035' (IEA, 2011: 2). This illustrates the urgent need to increase renewable energy in the overall electricity supply. This is easier said than done, however, because a main cause of the looming lock-in challenge is that, historically, renewables have not been perceived to be cost competitive, when compared with fossil-fuel-based electricity generation (Arent et al., 2011; Haas et al., 2004).

Recently, there has been progress towards increasing renewable energy's contribution to the overall global energy supply. In 2010 an estimated 19.4 per cent of overall electricity consumption was supplied by renewable energy (16.1 per cent hydro and 3.3 per cent other renewables), with nearly half the 2010 added electricity capacity of approximately 194 gigawatts (GW) coming from renewable sources (39 GW Wind, 30 GW Hydro, 17 GW Solar PV) (Janet and Martinot, 2011). The contribution of renewables towards added capacity is particularly important, because worldwide demand for energy continues to increase, particularly in emerging economies such as China, India and South Africa.

In order to avoid locking in traditional fossil fuel systems, alternative energy sources must become a major contributor to meeting the future energy demand of developing economies. The global centres of growth have shifted in the twenty-first century, which has important implications for the geographical focus of renewable energy MBIs. The Organisation for Economic Co-operation and Development (OECD) countries are no longer the drivers, as 'non-OECD countries account for 90 per cent of the energy demand growth over the period from 2010–2035' (IEA, 2011). Such growth trends highlight the need for effective renewable energy promotion policies that address the challenges inherent in developing countries with limited statehood.

While the potential of MBIs to deliver cost-effective environmental policies has been widely recognised by policymakers, the focus of MBI theory has largely been placed on OECD countries (Stavins, 2002; Sandor et al., 2002; Aldy and Stavins, 2012). As a consequence, limited research exists on the applicability of MBIs in areas of limited statehood, particularly developing countries where the state capacity to fully implement and enforce policy mandates is often lacking (Börzel and Risse, 2010). As pointed out by Börzel et al. (2010), even in South Africa where legal standards and regulation are quite advanced, capacity for implementation and compliance is often lacking.

To overcome the challenges to renewable energy deployment, a policy shift supported by effective incentive mechanisms is needed to catalyse a rapid transition towards renewable energy sources (Christensen et al., 2006). A number of approaches to promoting renewable energy have been implemented, each requiring significant involvement from the state. The ongoing debate over what approach is the most successful and effective, focuses on price-driven versus demand-driven strategies (Haas et al., 2004). Price-driven strategies are characterised by the state setting a price that is intended to reduce uncertainty and attract IPPs. The best known mechanism is the feed in tariff (FIT), whereby the government sets premium tariff prices through long-term contracts for electricity from renewable energy generated by IPPs (Huang and Wu, 2011). On the other side of the debate is the demand-driven approach.

Within this strategy, the state sets an objective to be reached, typically using quotas on electricity suppliers through a system of tradable green certificates (Menanteau et al., 2003).

Fixed-price FITs have been used effectively to exceed renewable energy targets in Germany (Janet and Martinot, 2011). However, they have been criticised as not being cost efficient (Frondel et al., 2010; Butler and Neuhoff, 2008; Krewitt and Nitsch, 2003). To address this, the German Renewable Energy Sources Act is periodically reviewed, and FIT tariffs are revised to align with the maturing renewable industry and with decreasing installation costs. A fixed-price FIT was initially planned for South Africa (NERSA, 2009). However, this was abandoned in favour of a process in which FIT prices are determined through staged competitive bidding that began in August 2011.

Furthermore, an enabling environment must be created within local, national or international structures that can generate sufficient incentive for IPPs to engage at a project level (Dunkerley, 1995). Such incentives for promoting renewable energy should 'offer a reasonable risk-return ratio to investors and ... minimise the total costs for society' (von Flotow and Friebe, 2011). To address both these requirements effectively, MBIs develop in a co-evolutionary manner. The IPPs' need for financial incentive, and the state's willingness and ability to subsidise, are what make the co-evolutionary relationship possible. The desired outcome is a well-designed MBI emerging through co-evolutionary interactions between IPPs and the state, which culminate in incentives set to entice private sector participation at the lowest cost to society.

The players within government will include, at a minimum, the department of energy, the state energy regulator, the utility that supplies and/or distributes the electricity via the power grid (or multiple utility companies when deregulation has occurred) and the local municipal government. On the other hand, the renewable energy private sector includes a range of organisations that come together to form or support IPPs, such as project developers, equipment manufacturers, debt and equity providers and the landowners. Figure 4.1 is a high-level illustration of the key actors in the co-evolutionary framework.

The IPP is dependent on the state to ensure a sufficient return on investment for its renewable energy projects, and the state is dependent on the IPP to deliver clean energy to the country's electricity supply. Due to limited resources in developing countries, any additional costs associated with promoting renewable energy must be fully understood and evaluated within the context of all other immediate development challenges facing the host country.

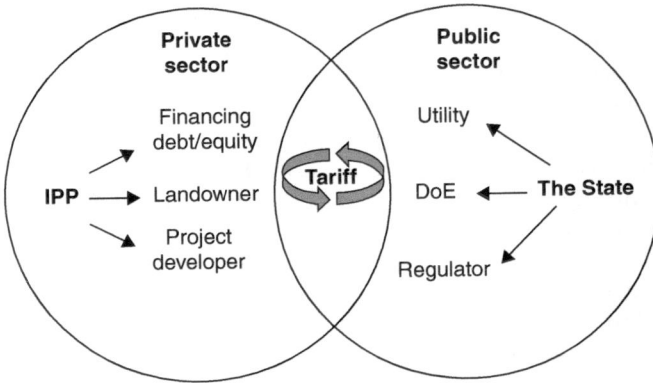

Figure 4.1   Co-evolution of renewable energy incentive design

The private sector is primarily concerned with achieving their expected rate of return for the project, which is based on the underlying risk profile of the venture. It has also been acknowledged that non-economic factors (e.g. administrative hurdles, grid access procedures, non-transparency) influence decision making by project developers (Lüthi and Prässler, 2011). However, assuming sufficient natural resources exist, and authorisation is feasible, the main concern of the business organisation is an acceptable financial return, in line with the risk profile of the project and host country. As a result the IPP is motivated by the business opportunity established by the state and is incentivised to engage actively with defining the rules and regulation that result in an attractive MBI (Toke, 2007; Dinica, 2006).

Considerations of renewable energy promotion in areas of limited statehood

Effective MBIs in developing countries must be cognisant of the host country's level of statehood and the corresponding perceived risk. In developing countries, MBIs act in part as the functional equivalent of the shadow of hierarchy cast by the state (Börzel and Risse, 2010). MBIs are required to bridge the cost differential between traditional fuel sources and renewables, and to provide assurance that IPPs will receive a financial return that corresponds with the host countries' risk.

   The financial support of the tariff price is a result of the current *tariff price differential* between fossil-fuel-based power generation and renewable energy. The key challenge is to determine a price per kilowatt hour (kwh) that can provide an appropriate private sector return on investment. The desired outcome is a tariff price that represents the lowest possible cost of renewable energy but is still attractive to IPPs. Key considerations include the following: What is the differential cost between renewable energy cost and the current tariff rate? Who will pay the differential between the existing electricity price and the cost of renewable energy? What is the price sensitivity of the end electricity consumer? These questions have different answers depending on the country's context and level of statehood. Within South Africa, even though climate change threatens to have a disproportionally negative impact on Africa (Stern, 2007; Parry, 2009), paying higher costs for electricity is often viewed as unacceptable, given the numerous and more urgent development challenges facing the country (Fay et al., 2011; Vorster et al., 2011).

The private sector has an entirely different set of drivers compared to the state. Classic risk-versus-reward analysis is invoked, with the level of statehood as a key determinant of risk. Moving from developed to developing countries has significant overall cost implications because investors demand an increased internal rate of return[1] (IRR) in areas with a lower level of statehood. Renewable energy projects are highly dependent on financing terms, with the IRR determined in large part by the risk associated with anticipated cash flows (Wiser and Pickle, 1998). This means that IPPs must have confidence that the host-country government will honour contractual obligations and continually fund the subsidy over the long term.

Global capital has become a potential resource for developing countries, and power projects are often dependent on global financing for investment. Investors prefer investing in countries with manageable country and political risks. They will only invest in higher-risk projects and countries if they receive a higher return compared to lower-risk investment options in developed countries. The viability of any renewable energy project is determined by its ability to achieve a financial return in excess of the perceived risk associated with the project. IPPs decide on projects based on their potential to achieve the equity IRR, which is determined by the risk profile of the project and host country.

A recent survey conducted by von Flotow and Friebe, exploring investor's preferences for different framework conditions for investments in wind parks in emerging and developing countries, concluded that:

> Investors want to reduce political investment risks (framework conditions) as much as possible. An increased risk in comparison to an

ideal scenario can cause investors to hold back their investment or at least raise their return expectations correspondingly in order to compensate for the increased risk. (2011: 19)

This runs counter to the need for the governance system to minimise the overall cost of electricity. As a result, it becomes more difficult to make a compelling financial case for promoting renewable energy in developing countries, and this underscores the importance of effective MBI design. In the next section, a comparison between Germany and South Africa highlights how perceived level of statehood affects the financing of renewable projects.

Germany and South Africa wind energy projects in comparative analysis

Due in large part to level of statehood, there are wide variations in the cost of wind energy projects across countries (Schwabe et al., 2011). In an effort to elucidate these profound differences, a comparative analysis of wind energy in Germany and South Africa is provided. First, the expected IRR benchmark is estimated for South Africa and Germany. Then, using the estimated IRR benchmarks, the tariff price required to meet this investment hurdle is modelled using hypothetical wind farms in both countries.

Estimating the cost of capital for renewable energy projects is subjective, with little consensus on the preferred approach. The main options for estimating cost of capital are: 1) to use historical market data (ex-post) using the capital asset pricing model (CAPM); 2) to estimate future cash flows (ex-ante) to determine rate of return of equity using the discounted cash flow method. For the purpose of this analysis, CAPM using historic market data is used, because it incorporates the risk attributes specific to the country, project and industry. South Africa and Germany both have well-functioning financial systems with decades of reliable market data to draw upon. Furthermore, research has shown that this approach is preferred in practice. Bruner et al. (1998) researched the leading firms identified in a report, *Creating World Class Financial Management: Strategies of 50 Leading Companies,* for their excellence in strategic financial management. They conclude that these firms prefer CAPM to estimate the cost of equity. Furthermore, Kaplan and Peterson (1998) suggest that the CAPM has a strong theoretical foundation and simplicity, one of the reasons for its widespread use in practice.

In the case of renewable energy projects, equity investors have to consider many factors, including the economic, political, legal, and industry conditions of the host country. The proposed model to estimate required IRR is a market-based model where the return on a broad index of securities acts as a valid proxy for these common macroeconomic factors. This represents the market observations, based on the enduring economic relationships between elements of the real economy, which can be viewed as an approximate measure of statehood. Specifically, the market proxies used for estimating the equity IRR are yield to maturity (YTM) on long-term government bonds and the equity risk premium.

Furthermore, wind power is new to South Africa. The first round of negotiations for power purchase agreements for renewable IPPs was announced in December 2011 as part of the competitive bidding process. The lack of South African experience with wind power causes further uncertainty for investors relating to the wind resource, construction, operation, and maintenance of turbines. As a result, a new industry risk premium is applied by investors to compensate for the nascent stage of the wind power sector in South Africa. Over time, however, as South Africa gains experience in wind power, the new industry variable will decrease to zero.

The CAPM approach to estimate the equity IRR for renewable energy projects has been developed in previous work by Fay and Kumar (2011). The model allows the investors to be compensated in two ways: time value of money and risk. The estimates are market driven and, therefore, are an objective consensus of all investors in the financial market. The proposed model to estimate the cost of capital for renewable energy projects is as follows:

Expected return on equity $E(R_e)$ = YTM on a long term government bond
+ equity risk premium
+ new industry risk premium

Applying the model to South Africa as of January 2012 yields an expected rate of return of 20.77 per cent:

South African $E(R_e)$ = 8.45% + 7.29% + 5.03% = 20.77%

The YTM on the Long Term Government Bond rate of 8.45 per cent is reported from the Johannesburg Stock Exchange's GOVI Index as of 3 January 2012. The 10-year equity risk premium of 7.29 per cent is taken from Hassan and Van Biljon's analysis (2010). The new industry premium of 5.03 per cent is calculated using the Ibbotson Associates full-information beta estimation process.[2]

By comparison, Germany, with a higher perceived level of statehood and a more advanced wind sector than South Africa, benefits from a much lower expected rate of return at 7.30 per cent:

$$Germany\ E(R_e) = 1.90\% + 5.40\% + 0\% = 7.30\%$$

The YTM on the long-term government bond rate of 1.90 per cent is from 3 January 2012, and is based on the German-government-backed 10-year bond (Bloomberg, 2012). The equity risk premium of 5.40 per cent is estimated via an in-depth survey approach performed by Fernandez, Aguirreamalloa and Corres (2011). Germany has one of the most developed renewable energy sectors in the world. It is a leader in the manufacture of wind turbines and had over 27.2 GW of wind operating at the end of 2010 (Janet and Martinot, 2011: 20). Due to the advanced state of the wind sector in Germany, the new industry risk premium of zero is applied.

The IRR benchmarks demonstrate a significant difference between cost of capital in Germany and South Africa. Germany, with a high level of statehood and an advanced wind energy sector, benefits from a low cost of capital for the wind projects. Conversely, South Africa has a lower level of statehood combined with a nascent wind energy sector, and this results in a higher perceived risk that in turn demands a higher cost of capital.

Next, the effect of cost of capital on the overall tariff price is explored for Germany and South Africa. A financial model is developed for hypothetical 100 MW wind farms in both countries in order to estimate the tariff price required to meet the projects expected return on equity. The purpose of this analysis is to illustrate the overall price (as measured by required tariff) variation between comparable renewable power generation projects. Table 4.1 provides a summary of the key variables input into the financial model for the hypothetical South African and German wind farms. Indicative quotes from actual operating and maintenance service providers are used uniformly across both projects. All other uniform assumptions are wind industry norms or estimates from best available data. The lifespan of both projects is 20 years. The key variable inputs are capital expenditure per MW wind installed, capacity factor, consumer price index (CPI) and the interest rate for debt financing.

For the South African model, the Standard Bank benchmark of US$2,000,000 per MW installed is used to estimate the cost of building wind farms in South Africa (Standard Bank, 2011). The nine per cent interest rate used is the South African prime-lending rate as of

Table 4.1 Hypothetical wind farm financial modelling assumptions

Technical	South Africa	Germany	Unit
Wind turbine capacity	2.5	2.5	MW
Turbines	40	40	Turbines
Total capacity	100.0	100.0	MW
Wind capacity factor	27.17%	25.80%	%
Estimated annual net generation	238.009	226.008	MWh

Operating	Value	Value	Unit
O&M cost annual (indexed by inflation)	29.66	29.66	Million Rand
Socio-economic development Fee	2.1%	0.0%	% Revenue
Annual escalation in tariff (CPI)	5.0%	2.3%	% Increase
Inflation (CPI)	5.0%	2.3%	%
Land rental fee	1.5%	1.5%	% Revenue
Salvage value	5%	5%	% of turbine cost

Financial	Value	Value	Unit
Capital expenditure	R 1,473.71	R 1,399.16	Million Rand
Debt	70.00%	70.00%	%
Equity	30.00%	30.00%	%
Equity IRR	20.77%	7.30%	%
Interest rate – debt	9.00%	3.58%	%
Loan repayment tenure	15	15	Years
Tax rate	28.00%	28.00%	%
Exchange rate	1 Rand	10 Euro	Rand to Euro

January 2012, and the CPI is the 2011 average of five per cent (Statistics SA, 2012). The potential wind resource uses a capacity factor of 27.17 per cent as estimated by a capacity study that relates to the first 2,000 MW of installed wind capacity in South Africa (Werner et al., 2011).

While capital expenditure outlays have varied over the years by project, an average German specific benchmark of €1,373,000 per MW installed, determined by the International Energy Agency is used (Schwabe et al., 2011). The interest rate of 3.58 per cent is sourced from Bundesbank, using the rate that the domestic banks in Germany charge on euro-denominated loans to non-financial corporations domiciled in the Euro area as of November 2011. The inflation rate used is Germany's 2011 average CPI of 2.32 per cent (CPI, 2011). The estimated average capacity factor for Germany is 25.8 per cent, based on 2008 wind farm performance results (Schwabe et al., 2011).

The above analysis results in a rough estimation of the inflation indexed tariff rate necessary over a 20-year period for IPPs to engage in South Africa and Germany. A tariff rate of €6.3c per kwh is necessary to meet the expected return on equity of 7.30 per cent in Germany, while a tariff rate of €9.7c is necessary to meet the expected return on equity of 20.77 per cent in South Africa. This represents a 54 per cent higher cost to generate wind energy in South Africa. Such a profound price differential, even with the better-anticipated wind resource in South Africa, underscores the implications of the host-country risk on financing. The financial engineering of renewable energy projects further compounds the difficulties, because of the substantial capital needed up front, and the long-term repayment, based on the power purchase agreement. As a result, cost of capital plays a larger role in the viability of renewable energy projects in countries with higher levels of limited statehood; and any incentive mechanism must understand and address each country's specific cost of capital situation.

From the perspective of the state, it is important to understand the cost differential between renewable energy and the current tariff price. Even though direct comparisons of IPP generation costs and existing tariffs from national utilities are difficult to fully discern (Eberhard and Gratwick, 2011), they are useful for providing a basic understanding of the price differential, when including new energy sources in an incumbent system. Table 4.2 compares the hypothetical tariff rate required by wind energy IPPs, the current retail household kWh tariff rate and the on-shore wind tariff rate offered in the two countries. What emerges from the German case is that the retail price of €26.7c is significantly higher than the estimated cost of €6.3c for wind energy generation. This is not a straight comparison, because the overall tariffs in Germany are comprised of numerous taxes and charges, and thus exact cost of generation can only be estimated. For example, an analysis of the 2005 electricity tariff in Germany found that 60 per cent of the tariff contributed to electricity generation, transmission and marketing, 10 per cent to a concession charge and the remaining 30 per cent towards taxes (Wenzel, 2006).

To estimate the actual cost of electricity generation for Germany, the average base price for electricity on the European Power Exchange Spot market for Germany/Austria was €51.12 per MWh in 2011, which corresponds to approximately €5.1c per kWh (EPEX Spot 2012). The FIT pricing used in Germany shows that the estimated generation cost for

Table 4.2   Tariff pricing comparison

Country	Estimated wind generation cost	Household tariff kWh	Wind energy pricing
Germany	6.3c[3]	26.7c*	5c – 9c***
South Africa	9.7c	6.07c**	11.5c****

Euro cent denominated.

*June 2011 retail household end-user with consumption of 3 500 kWh/year, source: http://www.energy.eu/.

**2012/2013 standard average price, source: http://www.eskom.co.za/c/53/tariffs-and-charges/.

***Estimated German FIT pricing 2011, source: http://www.energy.eu/#Feedin.

****Maximum bid allowed under REBID for on-shore wind energy is 115c ZAR.

2011 electricity in the German grid network is only slightly above the hypothetical cost for wind power and at the low end of the German FIT. This implies that wind energy in Germany is getting close to being cost-competitive as opposed to traditional generation.

In direct contrast, the South African state has fewer options available to promote renewable energy, because the gap in price differential between wind and the retail electricity price is significant. ESKOM, the monopolistic South African utility, currently provides electricity to households at a retail price of approximately €6.07c (NERSA, 2012). This tariff price includes increases over the past three years of approximately 25 per cent in 2010 and 2011 and 16 per cent in 2012 (ESKOM, 2011). However, even with these increases, the cost of generating coal-based power, using South Africa's legacy infrastructure, is still comparatively inexpensive (Pegels, 2010). This presents a challenge to fund the differential between the current tariff and the estimated €9.7c to generate wind energy in South Africa.

Another important consideration that underpins this analysis is the willingness and ability of the end electricity consumer to absorb higher costs for electricity. As is the case with most developing countries, South Africa has high levels of poverty and inequality, resulting in substantial price sensitivity for a large part of the population to tariff price increases. Compounding the problem is the fact that the government has limited resources and must prioritise the most immediate needs, such as health and education, to name a few. As a result, subsidy funding for renewable energy is a secondary concern, because the country faces a host of developmental challenges that are perceived as more important and immediate than climate change (Winkler and Marquand, 2009). On the other hand, Germany has a much higher electricity price and a populace less sensitive to price increases. There has also been a consistent willingness, by the German consumer, to buy renewable energy at a higher price (Wüstenhagen and Bilharz, 2006), providing much more flexibility and support for renewables in Germany.

Discussion: MBI policy in areas of limited statehood

The energy sector is dynamic and constantly changing, with a number of trends supporting renewable energy deployment. The overall costing trend for installing renewable capacity has been decreasing each year, as technology improves and greater scale is achieved. Furthermore, renewables generally reduce the overall volatility of energy project pricing structures because there are no fuel costs. Wind energy investors have fairly certain knowledge of the lifetime cost of the plant from the outset, because installed costs and mean wind speed are known, and there are low variable costs, zero fuel cost, and no carbon emission costs (Schwabe et al., 2011).

On the other hand, underlying fuel costs of traditional energy sources (i e. coal, oil, gas, nuclear) are highly variable and increasing, because they are subject to the world market and the increasing demand for energy. The cost differential between renewables and fossil-fuel-based electricity is also highly country specific, because both renewable energy potential and availability of fossil fuels differ from country to country. Furthermore, developing countries are often characterised by dominant state-owned utilities that are 'rarely exposed to market costs of capital' (Eberhard and Gratwick, 2011: 5542). This manifests in widely varying price differentials between the current electricity tariff offered by incumbent national utilities and the price required by IPPs to provide electricity from renewable sources profitably.

For renewable energy in developing countries to reach the scale required to meet the 450 scenario, MBI design must consider carefully both cost of capital and the potential role of incentives in decreasing project risk. By doing so, MBIs will increase their cost efficiency and shorten the time-frame until renewable electricity generation reaches parity with traditional energy systems. This will require innovative approaches to incentivise both the state and IPPs.

Recommendations include implementing mechanisms that provide below-market-rate loans and loan guarantees to the IPP, in order to reduce risk and drive down the cost of capital. The host countries' level of statehood could be the determining factor if the mechanism is managed by the international climate regime or the national government. Countries with limited statehood would then benefit from an international mechanism, which could draw down the project's host-country risk. Precedent exists for such mechanisms at the national level. For example, the German government has successfully provided below-market-rate loans to support its renewable energy sector. As early as the 1990s, wind energy plants and PV roof systems were eligible for soft loans with an interest rate reduction of 4.5 per cent as compared to standard loans (Paolo, 2006; Rosaria Di Nucci et al., 2007). In 1999, Germany introduced the market incentive programme (MAP) for smaller-scale renewable energy systems that provided direct investment subsidies and soft loans with long term repayment conditions and partial debt forgiveness if certain conditions were met (Bechberger and Reiche, 2004). Recently, the South African government even began exploring such options through an initiative called South African Renewables Initiative (SARI), a collaboration with global donors and foreign governments to explore innovative funding mechanisms to lower the cost of renewable energy deployment (Creamer, 2011).

Another recommendation is to address the tariff price differential with mechanisms specifically designed to leverage the stable cost of wind energy versus the increasingly volatile fossil-fuel-based electricity costs. Theoretically, an international financing mechanism could engage directly with South Africa to provide immediate financial support to meet the incremental tariff price difference for wind energy. The business-as-usual tariff price could be marked to the predominant fuel source used in the host country, for example coal in South Africa. If designed effectively, opportunity may exist for funds to be repaid if and when wind energy becomes cheaper than coal-based electricity generation. In theory, the country that receives the initial financial payment to cover the present day negative tariff price differential could eventually be paid back by a future positive tariff price differential. The global coal price could be used as the benchmark to assess the traditional cost of power. This approach could allow developing countries to benefit immediately from inclusion of renewable energy, without having to absorb the present day higher cost. This is a reverse lock-in strategy, because South Africa would secure the power it needs to grow the economy via renewable energy at the same cost as coal-based generation.

Figure 4.2 illustrates the expected tariff pricing in South Africa versus an estimated wind energy tariff from 2012 to 2032. This demonstrates the potential competitiveness of wind generation over the long term. The South African overall tariff price is based on ESKOM's multi-year

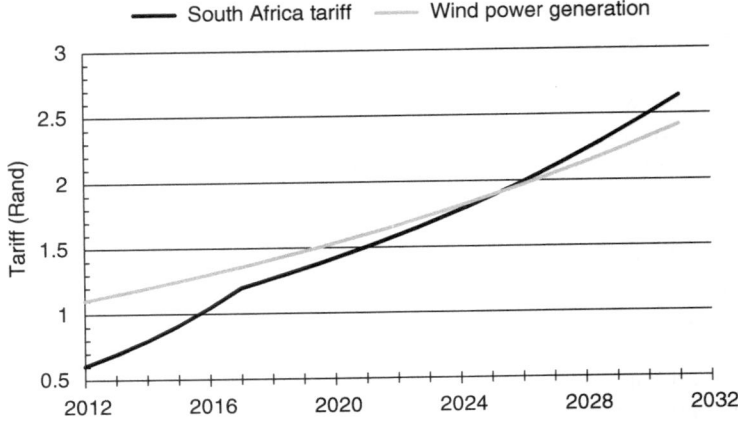

Figure 4.2   South Africa electricity tariff vs estimated wind power tariff

tariff pricing proposal until 2018 (de Lange, 2012), then reverting to a historical average of CPI from 1997 to 2012 of 5.79 per cent per annum (ESKOM, 2012). The wind energy tariff is stable, as it is set by the competitive bidding guidance from NERSA at 4.2 per cent (DOE, 2011) and assumes a 14 per cent mark-up to the estimated generation costs to include distribution and marketing (Wenzel, 2006).

This incentive mechanism is a long-term, speculative proposition, dependent on the variable cost of fossil-fuel-based electricity versus the stable cost of wind energy. While risky from a purely financial perspective, such an approach could be a cost-efficient and effective way for the international climate change regime to promote the global public good of reducing carbon emissions. Such a mechanism offers a sustainable alternative to fossil-fuel-based power generation, while at the same time supporting climate smart development because it does not require the host-country government to divert its limited resources to finance the more expensive current cost of renewable energy.

Conclusion

In order to meet the challenge posed by climate change, the current dependence on fossil-fuel-based electricity generation needs to undergo dramatic transformation in the immediate future. Otherwise, we risk locking in carbon-intensive power installations that will push the world past the 450 ppm threshold. This calls for bold climate change governance, in which the public and private sectors must work effectively together through a co-evolutionary framework to establish and implement effective incentives for renewable energy that are customised to the host countries' level of statehood.

Global climate change governance needs to place special focus on developing countries, which are the future demand centres for electricity. To do so, challenges that are inherent to areas of limited statehood need to be better understood and addressed. MBI approaches from the developed world may not work well in developing countries because host-country contexts are different. The host-country level of statehood is a key determinant of any renewable energy project's risk profile, manifesting in significantly higher required rate of return for South Africa as compared to Germany, irrespective of the new industry risk. This in turn influences the cost of capital. The result is renewable energy that is significantly more expensive in the developing world than in OECD countries, expanding the tariff price differential and hindering the state's ability to prioritise the large-scale promotion of renewables. To overcome this challenge, international and national MBIs are needed to reduce the required cost of capital arising from developing countries lower level of statehood.

Notes

1. Internal rate of return is the metric used to measure and compare the profitability of potential investments.
2. IRPi = (RIi x ERP) − ERP; where IRPi indicates the expected industry risk premium for industry i; RIi is the risk index (full information industry i beta) estimated using the top ten alternative energy exchange-traded funds as the proxy of new industry risk premium (beta =1.69); and ERP represents the expected equity risk premium.
3. Exchange rate of 10 to 1 is utilised to convert from Rand to Euro.

References

Aldy, J.E. and Stavins, R.N. (2012) 'Using the Market to Address Climate Change: Insights from Theory and Experience', *Daedalu* 141, 2, 45–60.

Arent, D.J., Wise, A. and Gelman, R. (2011) 'The Status and Prospects of Renewable Energy for Combating Global Warming', *Energy Economics* 33, 4, 584–593.

Bechberger, M. and Reiche, D. (2004) 'Renewable Energy Policy in Germany: Pioneering and Exemplary Regulations', *Energy for Sustainable Development* 8, 1, 47–57.

Bloomberg (2012) Germany Government Bond Yield and Interest Rates, 7 January 2012.

Börzel, T.A., Héritier, A., Kranz, N. and Thauer, C. (2010) 'Racing to the Top?' *in Governance without a State* (New York: Columbia University Press).

Börzel, T.A. and Risse, T. (2010) 'Governance without a State: Can it work?', *Regulation and Governance* 4, 113–134.

Bruner, R.F., Eades, K.M., Harris, R.S. and Higgins, R.C. (1998) 'Best Practices in Estimating the Cost of Capital: Survey and Synthesis', *Financial Practice and Education* Spring/Summer, 13–28.

Butler, L. and Neuhoff K. (2008) 'Comparison of Feed-In Tariff, Quota and Auction Mechanisms to Support Wind Power Development', *Renewable Energy* 33, 8, 1854–1867.

Christensen, J., Denton, F., Garg, A., Kamel, S., Pacudan, R. and Usher, E. (2006) *Changing Climates: The Role of Renewable Energy in a Carbon-Constrained World* New York: United Nations Environment Programme).

CH (2011) Inflation Germany 2011. Available at: http://www.inflation.eu/inflation-rates/germany/historic-inflation/cpi-inflation-germany-2011.aspx.

Creamer, T. (2011) 'Renewables Procurement Process May Be Extended Beyond the First Phase', *Engineering News* 16 December 2011.

de Lange, J. (2012) 'Industry Seeks Talks over Eskom Tariff', *BDO South Africa* 16 July.

Department of Energy, Republic of South Africa (2011) Renewable Energy IPP Procurement Programme: Department of Energy Renewable Energy Bidders Conference. 14 September 2011.

Dinica, V. (2006) 'Support Systems for the Diffusion of Renewable Energy Technologies – An Investor Perspective', *Energy Policy* 34, 4, 461–480.

Dunkerley, J. (1995) 'Financing the Energy Sector in Developing Countries: Context and Overview', *Energy policy* 23, 11, 929–939.

Eberhard, A. and Gratwick, K.N. (2011) 'IPPs in Sub-Saharan Africa: Determinants of Success', *Energy Policy* 39, 9, 5541–5549.

Eisenhardt, K.M. (1989) 'Building Theories from Case Study Research', *The Academy of Management Review* 14, 4, 532–550.

Eisenhardt, K.M. and Graebner, M.E. (2007) 'Theory Building from Cases: Opportunities and Challenges', *Academy of Management Journal* 50, 1, 25–32.

EPEX Spot (2012) EPEX Spot. *European Power Exchange*. 01 February.

Eskom (2011) ESKOM RETAIL TARIFF ADJUSTMENT FOR 2011/12. Eskom.

Eskom (2012) Average Price Increases. Eskom Tariffs and Charges. URL: http://www.eskom.co.za/c/article/143/average-price-increases/.

Fay, J. and Kumar, U. (2011) 'A Market Model Based Financial Benchmark Required to Access Finance for Renewable Energy Projects in South Africa', *Green Economics: Theory and Practice presented at the AESS Annual Meeting 2011*, Burlington, Vermont, USA.

Fay, J., Kapfudzaruwa, F., Na, L. and Matheson, S. (2011) 'A Comparative Policy Analysis of the Clean Development Mechanism in South Africa and China', *Climate and Development* 4, 1, 40–53.

Fernandez, P., Aguirreamalloa, J. and Corres, L. (2011) *Market Risk Premium Used in 56 Countries in 2011: A survey with 6014 Answers*, IESE Business School.

von Flotow, P. and Friebe, C. (2011) *Framework Conditions for Investments in Wind Parks in Emerging and Developing Markets the Investors' Perspective* (Oestrich-Winkel: Sustainable Business Institute).

Frondel, M., Ritter, N., Schmidt, C.M. and Vance, C. (2010) 'Economic Impacts from the Promotion of Renewable Energy Technologies: The German Experience', *Energy Policy* 38, 8, 4048–4056.

Haas, R., Eichhammer, W., Huber, C., Langniss, O., Lorenzoni, A., Madlener, R., Menanteau, P., Morthorst, P.-E., Martins, A., Oniszk, A., Schleich, J., Smith, A., Vass, Z. and Verbruggen, A. (2004) 'How to Promote Renewable Energy Systems Successfully and Effectively', *Energy Policy* 32, 6, 833–839.

Hassan, S. and Van Biljon, A. (2010) 'The Equity Premium and Risk-Free Rate Puzzles in Turbulent Economy: Evidence from the 105 Year of Data from South Africa', *South African Journal of Economics* 78, 1, 23–39.

Huang, Y.H. and Wu, J.H. (2011) 'Assessment of the Feed-in Tariff mechanism for Renewable Energies in Taiwan', *Energy Policy* 39, 12, 8106–8115.

IEA (2008) World Energy Outlook 2008 (Paris: International Energy Agency).

IEA (2011) World Energy Outlook 2011 (Paris: International Energy Agency).

Janet, S. and Martinot, E. (2011) *Renewables 2011: Global Status Report*, Paris: REN21.

Kaplan, P.D. and Peterson, J.D. (1998) 'Full-information Industry Beta', *Financial Management* 27, 2, 85–93.

Krewitt, W. and Nitsch, J. (2003) 'The German Renewable Energy Sources Act—an Investment into the Future Pays off already Today', *Renewable Energy* 28, 4, 533–542.

Lüthi, S. and Prässler, T. (2011) 'Analyzing Policy Support Instruments and Regulatory Risk Factors for Wind Energy Deployment – A Developers' Perspective', *Energy Policy* 39, 4876–4892.

Menanteau, P., Finon, D. and Lamy, M.L. (2003) 'Prices versus Quantities: Choosing Policies for Promoting the Development of Renewable Energy', *Energy Policy* 31, 8, 799–812.

NERSA (2009) Renewable Energy Feed In Tariff. National Energy Regulator of South Africa.

NERSA (2012) NERSA Review Eskom's Tariffs For The Period 01 April 2012 TO 31 March 2013. National Energy Regulator of South Africa. 09 March 2012.

Pachauri, R.K. and Reisinger, A. (eds) (2007) 'Climate Change 2007: Synthesis Report', *Contribution of Working Groups I, II and III to the Fourth Assessment Report of the Intergovernmental Panel on Climate Change*, IPCC.

Paolo, A. (2006) 'Use of Economic Instruments in the German Renewable Electricity Policy', *Energy Policy* 34, 18, 3538–3548.

Parry, M. (2009) Climate Change is a Development Issue, and Only Sustainable Development can Confront the Challenge, *Climate and Development* 1, 1, 5–9.

Pegels, A. (2010) 'Renewable Energy in South Africa: Potentials, Barriers and Options for Support', *Energy Policy* 38, 9, 4945–4954.

Rosaria Di Nucci, M., Mez, L., Reiche, D. and Bechberger, M. (2007) 'Intelligent Energy Europe: Country Report Germany (Workpackage 3 No. 2)', Freie Universitat Berlin: Environmental Policy Research Centre-FFU, Berlin.

Sandor, R.L., Bettelheim, E.C. and Swingland, I.R. (2002) 'An Overview of a Free-market Approach to Climate Change and Conservation', *Philosophical Transactions of the Royal Society A: Mathematical, Physical and Engineering Sciences* 360, 1797, 1607–1620.

Schwabe, P., Schwabe, S. and Hand, M. (2011) IEA Wind Task 26: Multi-national Case Study of the Financial Cost of Wind Energy, Work Package 1 Final Report.

Standard Bank (2011) The IPPPP RFP. Standard Bank.

Statistics S.A. (2012) *Consumer Price Index (CPI), Past CPI Releases: January 2012*, South Africa: Statistics South Africa.

Stavins, R.N. (2002) 'Lessons from the American Experiment with Market-Based Environmental Policies', *SSRN Electronic Journal* 22, 01–032.

Stern, N. (2007) *The Economics of Climate Change: The Stern Review* (Cambridge UK: Cambridge University Press).

Toke, D. (2007) 'Renewable Financial Support Systems and Cost-effectiveness', *Journal of Cleaner Production* 15, 3, 280–287.

Vorster, S., Winkler, H. and Jooste, M. (2011) 'Mitigating Climate Change through Carbon Pricing: An Emerging Policy Debate in South Africa', *Climate and Development* 3, 242–258.

Wenzel, B. (2006) *What Electricity from Renewable Energies Costs* (Germany Federal Ministry for the Environment, Nature Conservation and Nuclear Safety).

Werner, D., Aphane, O., Otto, A. and Leask, K. (2011) Capacity Credit of Wind Generation in South Africa, Pretoria, South Africa: Deutsche Gesellschaft für Internationale Zusammenarbeit (GIZ) GmbH, Department of Energy, Eskom, Pretoria, South Africa.

Winkler, H. and Marquand, A. (2009) 'Changing Development Paths: From an Energy-intensive to Low-carbon Economy in South Africa', *Climate and Development* 1, 1, 47–65.

Wiser, R.H. and Pickle, S.J. (1998) 'Financing Investments in Renewable Energy: The Impacts of Policy Design', *Renewable and Sustainable Energy Reviews* 2, 4, 361–386.

Wüstenhagen, R. and Bilharz, M. (2006) 'Green Energy Market Development in Germany: Effective Public Policy and Emerging Customer Demand', *Energy Policy* 34, 13, 1681–1696.

5
Voluntary Collective Commitment: The Case of Business and Energy Efficiency in South Africa

Christopher Kaan and Stine Klapper

Introduction

Energy efficiency will remain one of the most pressing issues in energy policy in the future. The challenge to stabilise or even decrease the amount of energy consumed, while maintaining living standards, is one of the most important tasks for energy policy makers in developed countries. Developing countries, however, find themselves confronted with the need to catch up economically in order to improve their socio-economic situation. At the same time, warnings about the environmental effects of energy-intensive growth are becoming more pronounced, particularly with regard to the emission of greenhouse gases (GHGs) and associated climate change.

These connected challenges of economic development and GHG emissions create the opportunity for a decisive role for the business sector. A significant share of energy is consumed by companies. Because they are responsible for much of the emissions, companies can also contribute crucially to emissions reduction, by taking initiatives such as energy efficiency measures. Evidence shows that companies do not necessarily engage in competition for lower costs at the expense of social and environmental values, as the well-known 'race-to-the-bottom' thesis predicts. Numerous cases have been found of companies that implement higher environmental standards, as well as stronger measures for combating climate change, than are obligatory, and even agree collectively to reach defined targets (Prakash and Potoski, 2007; Flohr, Rieth et al., 2010). But why do companies make such commitments?

As an emerging market, South Africa faces the challenge of reconciling growth ambitions with environmental considerations. Its economy is 'inherently intensive' in energy use, due to the importance of the mining sector and the attached processing industry (DME, 2005). At the same time, natural resource abundance and the correspondingly low price for coal have lowered the significance of energy in production costs. Due to historically cheap electricity, the goal of energy efficiency has been less present in business strategies compared to other countries. The resulting high energy intensity has led to a high per capita greenhouse gas emission. South Africa ranks 45th, close to developed countries like the UK. Compared to the low GDP per capita (77), this is an enormous emission rate (DME, 2005; WBCSD, 2008).

In the last few years, energy efficiency has received increasing attention. The report *Energy policies for sustainable development in South Africa* identifies the improvement of energy efficiency as one of the key policy tasks for sustainable economic development in South Africa (Winkler et al., 2006). In accordance with this political goal, there is significant business commitment. The National Business Initiative (NBI) started its own energy efficiency and saving initiative, the Energy Efficiency Accord (EEA), in 2005. This voluntary agreement between businesses, in cooperation with the South African government, sets targets for a decrease in energy consumption, and implements corresponding reporting mechanisms. Hence, the EEA is a bi-partite co-regulation, since both public and business actors are involved. Additionally, it is a horizontal self-regulation via the business association NBI.

Energy production by coal is one of the major sources of GHG emissions, especially in South Africa. Energy consumption is therefore directly connected to GHG emissions, and the EEA can thus directly contribute to the mitigation of climate change. Energy efficiency measures are mainly a firm-internal activity. Yet, their effects – contribution to general emissions reduction and therefore climate change mitigation – have an impact that goes way beyond firm boundaries. Furthermore, the EEA itself gives various reasons for its existence. Among these are environmental concerns, and the NBI relates it explicitly to climate change mitigation (NBI and DME, 2005; Expert 1, 2012). Therefore, a general intention to contribute to the global public good of climate stability through mitigation measures is prominent in the EEA. The EEA creates a common framework with fixed reduction targets for all participants, providing higher standards than most companies might create on their own. Furthermore, it is a governance process through which commonly binding rules are established (Hamann and Börzel, Chapter 1). Irrespective of this background, individual drivers for firm participation may vary across the range of members. So why do companies in South African commit to the EEA?

This paper examines why businesses have decided to join the voluntary collective accord. Eight potential drivers are explored, all of which are all related to the theoretical framework of this volume (Hamann and Börzel, Chapter 1). It is not our intention to maintain a clear-cut distinction between these drivers. Rather, drivers are expected to appear simultaneously, and either to be mutually enforcing or to undermine each other. It is found that both institutional as well as organisational drivers matter: energy security, reputational concerns and prospects for information sharing and learning predominantly drive EEA membership. First, companies try to increase energy stability and support corresponding measures in the face of incapacities on the part of government to guarantee the provision of this collective good. Yet, concerns for energy security are the same among EEA members and non-members and can therefore not explain the different membership decisions. Second, the EEA is used to communicate the company's commitment to socialised norms, such as climate change mitigation, in order to gain a better reputation, which is an important economic asset. Third, companies use the EEA as an institutionalised vehicle for information and knowledge sharing in areas of uncertainty such as energy supply and climate change consequences.

These results are quite clear and show differences between participants and non-participants in voluntary collective commitments. Yet, the analysis also comes with a limitation. While the most influential drivers for EEA membership can be identified, further research is necessary to determine why certain drivers gain more momentum over some companies compared to others. For this, a comparative analysis for each individual driver is needed. This chapter provides grounds for this further research.

The following offers a short overview of the EEA and the South African energy supply situation. Following this, the theoretical assumptions are discussed, and will then be explored on the basis of the collected data.

The Energy Efficiency Accord

The EEA was signed in 2005 and reviewed in 2008. It is a voluntary agreement between business actors in coordination with the South African government. It was initiated to facilitate reaching the goals of the Energy Efficiency Strategy of the Republic of South Africa (DME, 2005), which stated that energy demand was to be reduced by 15 per cent while still allowing for economic growth. Importantly, the EEA membership is an explicitly voluntary commitment, lacking means of enforcement. Achieving higher economic growth remains the 'overarching imperative' (NBI and DME, 2005).

The EEA has been signed by various companies and the Ministry of Minerals and Energy. It is managed by the NBI, an alliance of both multinational corporations (MNCs) and South African national companies, and it focuses on sustainable development in South Africa. The NBI has been mainly responsible for collaborating with the government in drafting the EEA, in a process that involved stakeholder participation. The number of EEA signatories has risen from 30 members in 2005 to 42 companies, eight business associations and the Department of Energy, in 2011. The participating companies belong to different business sectors, including the most energy intensive, and South Africa's largest companies are involved (NBI and DME, 2005; WBCSD, 2008; NBI, 2011). To support EEA implementation, the Energy Efficiency Technical Committee was founded. The committee provides a forum in which EEA participants and stakeholders can meet and share ideas and best practices. It has been managed by the NBI, but has increasingly developed its own structures. This information exchange, including expertise and peer support, networks and reporting mechanisms, has been the main elements of the EEA. Another important element is the development of individual company strategies for energy efficiency. However, in recognition of the diverse firm contexts and circumstances, the EEA does not promote specific technologies and practices. The government expects assistance by the initiative for the development of effective regulations regarding energy efficiency issues. At the same time, the Department on Minerals and Energy promised prioritised information and leverage for the signatories concerning future revisions of the National Energy Efficiency Strategy and related regulatory modifications (NBI, 2011).

In 2011, the members replaced the EEA by the Energy Efficiency Leadership Network Pledge, which was introduced to renew the commitment to energy efficiency. This pledge was presented in the context of the COP17 in Durban and, like the EEA, involved government representatives, the NBI, companies and associations. The main elements are the development of a 'road map', internal energy efficiency targets, stronger reporting and capacity building (Mail and Guardian, 2011).

Applying the theoretical framework

The following presents theoretical considerations concerning why business leaders may decide to voluntarily commit to the accord.

All three institutional drivers described in the introduction are important to our case. Following the logic of a 'shadow of anarchy', the absence of regulations and the perception that future regulations are unlikely, may lead companies to develop their own initiatives to strengthen the institutional environment in which they operate. Businesses rely on common goods and regulations, which are necessary or beneficial for their functioning. One essential public good is energy supply. Businesses depend strongly on stable and secure availability of energy. Yet, South Africa has experienced serious energy insecurity, and the business sector has been affected by power shortages. In 2007 and 2008, in particular, shortages and blackouts hit the country. South Africa's public utility, Eskom, reacted to this by introducing a so-called 'load shedding' scheme, consisting of scheduled blackouts, which was used to secure grid stability. Plans have been developed to construct additional power plants and increase overall capacity. As a result, the historically low electricity price has risen significantly and will rise further in the future. Reducing overall energy consumption could improve supply stability and thereby the general business environment. Only collective efforts can lead to a substantial reduction of energy consumption, so government could be expected to provide such regulation. But governments in areas of limited statehood might lack the capacity to introduce and enforce such regulations (Börzel and Risse, 2010). If the state does not organise the provision of such a collective good, this can motivate proactive business behaviour. Thus, hypothesis 1 maintains that firms join and engage in the EEA to reduce costs and risks provoked by energy insecurity.

Another institutional driver is the 'shadow of hierarchy'. This concept describes the potential danger to business the state-imposed strict regulation that might result if the problem is not solved by voluntary self-regulation. These regulations could raise costs in the form of fines for non-compliance with certain rules. Importantly, these state sanctions can only be pre-empted if voluntary regulations involve sufficient actors. Yet, such a shadow of hierarchy depends on the strength of the institutions that potentially exercise control (Mayntz and Scharpf, 1995; Héritier and Lehmkuhl, 2008). Importantly, different companies' managements may have differing perceptions of the shadow of hierarchy and perceive the threat to different degrees. Hence, hypothesis 2 maintains that pre-empting coercive regulations by committing to a voluntary framework drives EEA membership.

The third institutional driver is seen in NGO activism and the influence of consumer preferences in the market. Ethically correct behaviour can be an economic asset and comparative advantage (Spar and LaMure, 2003; Smith, 2008). Correspondingly, ignoring norms can lead to reputation losses, which can strongly affect a company's economic performance (Haufler, 2001; Blanton and Blanton, 2007). Societal pressure, such as campaigning of transnational to local NGOs, can provoke consumer boycotts and hence lead to losses in market share (Newell, 2001; Flohr, Rieth et al., 2010). Participation in voluntary accords can be a way to underline and communicate a company's commitment and respect for a certain norm. This is especially relevant for preventing comparative advantages if competitors participate in the agreement. Although these kinds of 'image campaigns' may be judged more cosmetic than real (Porter and Kramer, 2006), they can still have substantive consequences. Reputation may also be relevant with respect to an entire sector. Preventing peers from spoiling the image of the sector may be a reason for pushing for a collective agreement (Hönke et al., 2008). Thus, hypothesis 3 says that reputation is the reason for EEA participation.

Organisational drivers can also be very relevant for companies' membership in the EEA. The first driver is culture and leadership commitment, which is directly affected by the degree to which normative considerations are taken into account in the business strategy. Thus, the desire to do what is considered 'right' or simply 'normal' can explain business actions. A company's management may therefore have decided to join the EEA because it considers energy saving and mitigating climate change as appropriate and 'right' behaviour. Pushing for a collective commitment may also be rooted in the motivation to combat climate change more effectively, since this can only be possible as a collective endeavour. Thus, hypothesis 4 sees the internalisation of norms, such as not harming the global climate, into the culture and leadership of the company as the reason for EEA commitment.

Norms also have economic cost–benefit implications (Hamann and Bözel, Chapter 1). First, if a norm-guided management that has decided to invest in energy efficiency measures expects high initial costs for this step, it might wish for the engagement of competitors, in order to avoid comparative economic disadvantages. Thus, fear of comparative economic disadvantage may be as strong a motivation for collective agreement as the wish to save the planet. Hence, hypothesis 5 expects norm-following companies to push for a collective agreement in order to avoid comparative disadvantages of high adjustment costs.

Another organisational driver for the initiative to adjust the individual energy scheme might be regulations originating from a company's home country. Internationally operating corporations very often have directives that affect all their companies worldwide and convey business systems and standards from their home countries. This is not necessarily an external shadow of hierarchy, however. It might well be more costly to allow for diverging practices in other countries if production procedures are streamlined and standardized across the entire international corporation (Jones, 1999; Murphy, 2000; Hall and Soskice, 2001). As a result, these companies might press for stricter regulations to avoid comparative disadvantages. If the government is unable to introduce these regulations, a self-imposed initiative can be a substitute. Thus, hypothesis 6 considers home country regulations of MNCs responsible for companies to push for a collective agreement on energy efficiency.

Organisations might follow the lead of other successful competitors. A further organisational driver for companies to join the EEA might simply be that 'everybody does it'. If a group of companies initiates and joins the EEA because of one of the drivers mentioned above, other companies might follow in imitation (Di Maggio and Powell, 1991; Bansal, 2005; Prakash and Potoski, 2006). Without understanding the complex relationships of issues such as energy efficiency and climate change, and the uncertainty in their developments, companies might just join the EEA because they fear negative economic consequences if they remain an outsider. Hypothesis 7 therefore addresses mimicry as the driver for EEA participation.

Finally, one important driver to make companies participate in a collective enterprise is the potential gain in information exchange and knowledge sharing between companies. As pointed out in hypothesis 7, when confronted with new and wicked challenges, companies stick to the successful strategies of their competitors. A collective agreement might help to increase the capacity of the companies to fulfil their goals regarding energy efficiency. Transparency among the members might help to select successful strategies and lower the transaction costs of the individual members (Chayes et al., 2000). Hypothesis 8 predicts that the information-sharing possibilities are a major reason for companies to join collective agreements.

As pointed out in Chapter 1, it is always a mix of various factors that construct preferences concerning the contributions of companies to provide a common good. In our case, a mixture of different motivations might lead to participation or non-participation in the EEA. Different factors might have caused the membership to participate to different degrees. The goal is therefore to identify what kind of patterns of reasoning exist.

To summarise, these eight drivers are assumed responsible for companies' involvement in the EEA: 1) shadow of anarchy, 2) shadow of hierarchy, 3) reputation, 4) culture and leadership commitment, 5) competitiveness considerations, 6) home country regulations, 7) mimicry, and 8) knowledge and learning. In the remainder of this chapter, we explore the explanatory power of these hypotheses.

In the next section, we will explore whether and to what extent the potential explanations are actually given by business representatives in interviews and documents. For this study, 12 interviews were conducted with individual company representatives, representatives of business associations and experts. Most of interviewees were mid-level management and responsible in the Corporate Social Responsibility (CSR) or Energy department of their companies. Further information was collected through an online questionnaire of some companies. In addition to the interviews and questionnaires, annual and sustainability reports were analysed.

The sample of the analysis comprises all members of the South African Chapter of the Energy Intensive Users Group (EIUG). The EIUG is a single-issue lobby group 'dedicated to the promotion of the interests of high quantity energy users in South African Industry. The EIUG aims to ensure the continued availability of low cost, high quality electricity supply on a reliable basis for energy intensive industrial players' (EIUG, 2012). The members are 'highly affected in view of energy intensity and dependence' and therefore 'firms in high-salience sectors' (Hamann and Börzel, Chapter 1, p. 15 in this volume). Interestingly, the membership in the EEA varies across EIUG members: only 12 of 33 participate in the EEA. The EIUG therefore provides an interesting and valid sample for the analysis

The drivers of the EEA

In this section, we present the findings for the eight drivers.

Shadow of anarchy

Energy security is an important issue and business risk for South African companies. Electricity blackouts have hit the industry in the past. The current instabilities affect operations, and future supply remains uncertain. This creates difficulties for ongoing production and hampers adequate planning. These difficulties were named by all companies, irrespective of their EEA participation or non-participation (Implats, 2011; Member 2, 2011; NBI, 2011; Member 4, 2012; Non-Member 1, 2012).

Similarly, incapacities on the part of the government and Eskom, South Africa's public utility, are perceived irrespective of EEA membership. It was generally considered unlikely that Eskom will be capable of providing a stable supply of electricity in the short and middle term. Although new power plants are being built – contributing to increasing electricity prices – it is not believed that they will function adequately and early enough (Member 3, 2012; Member 4, 2012).

The members of the accord see a connection between this energy insecurity and their own energy efficiency activities and EEA membership. This is partly presented as a general aim to become less vulnerable to energy volatility. Other members see a more direct link to the wish to help government fulfil its function and a clear intention to fill the gap left by government. One possible contribution is seen in the collective endeavour to increase general energy efficiency (NBI, 2011; Expert 1, 2012; Member 3, 2012; Member 4, 2012). Thus, one driver for participation in the EEA is the absence of the collective good of stable energy supply. The 'shadow of anarchy' drives companies' initiatives.

Although the EEA is seen as a support for stable electricity supply, non-members also pursue the same interest and show individual commitment. They do not participate in the accord but still engage in demand-management and offer support to government. The shadow of anarchy is widely perceived. Yet, they prefer individual commitments over collective agreements. They do not believe in the effectiveness of such a voluntary agreement. Thus, it is scepticism about the effectiveness of collective effort, rather than disapproval of the accord's intentions, which impedes the membership (Non-Member 1, 2012; Non-Member 2, 2012).

Thus, energy efficiency is a crucial driver for voluntary self-regulation and for contributing to providing the collective good of stable energy supply. Yet, since non-members also acknowledge the same need for action to secure energy supply, it does not determine why some companies decide to join while others do not. Energy insecurity is therefore not sufficient to explain EEA membership.

Shadow of hierarchy

Avoiding future sanctions cannot be considered the original driver for EEA participation since these were little perceived when the EEA was decided (Expert 1, 2012; Expert 2, 2012; Member 3, 2012).

There are different attitudes towards energy efficiency regulations. While some companies fear such regulations and prefer self-regulatory measures, others welcome regulation (Member 3, 2012; Member 4, 2012; Results from Survey, 2012). The reason for such support is the current uncertainty about energy policy. Some companies would prefer specific and binding targets, both in a government regulation and in agreements such as the EEA, as 'some sort of guidance', over individual activities in order to improve their planning security (Expert 1, 2012; Member 4, 2012). Others, however, want to avoid such provisions (Expert 1, 2012). In general, it is agreed that eventual regulations should reasonably weigh economic costs and environmental benefits (Member 3, 2012; Member 4, 2012). In summary, companies do not completely reject regulations. Yet, they want to avoid negative economic consequences of regulations. At the same time, they suffer from uncertainty regarding energy policy.

Perceived possible sanctions, however, were not the original driver for EEA membership. Although state regulations are now assumed likely to be developed, the general sentiment still reveals scepticism towards government capacity to enforce such regulations effectively (Hamann and Börzel, Chapter 1). Here, a distinction can be made between the planned carbon tax and an eventual regulation of energy efficiency. While effective enforcement of a carbon tax is expected, the effective implementation of energy efficiency regulations, which would be connected to government initiatives and engagement, are assumed less likely (Hamann and Börzel in this volume). The intent to agree on collective energy efficiency measures and to use the common forum of the EEA to influence government decisions on the eventual energy efficiency regulations is weak. Here, capacity deficits on the part of the government are also perceived in the process of policy development. Nevertheless, the aim to understand 'what's in the pipeline' is mentioned by business representatives, while other forums, such as Business Unity South Africa (BUSA) and EIUG, are considered as more important than the EEA or NBI context both by members and non-members (Expert 1, 2012; Expert 2, 2012; Member 3, 2012; Member 4, 2012; Non-Member 1, 2012).

Companies not participating in the EEA consider government regulations on energy efficiency less likely (Results from formalised interviews 2012), and expect incentive mechanisms rather than restrictions or sanctions (Non-Member 1, 2012). They nevertheless pronounce the same need for certainty as well as balanced policy choices, and they see a lack of government capacity in both policy development and implementation (Non-Member 1, 2012).

Therefore, a perceived 'shadow of hierarchy' or looming regulations can be considered an existent but very weak driver for EEA participation.

Reputation

Reputation has been a meaningful driver for EEA participation. EEA members state that energy efficiency has become a topic in public discourse, and that the public regards taking energy efficiency measures as a positive behaviour. All company representatives agree that the EEA functions as a public and 'visible' manifestation of commitment, and that it benefits reputation. This is regarded as the strongest effect of the EEA. Consequences for actual energy efficiency and climate change mitigation are considered much less substantial. Especially those producers related to the green technology market and to international high-end customers, state that they cannot afford a reputation loss; and this constitutes a 'hard business reason'. The relation between reputation and economic costs is therefore explicit, and it motivates public presentation of behaviour which is socially considered right (Sasol, 2006; NBI, 2007; Expert 3, 2012; Member 3, 2012; Member 4, 2012).

EEA membership is proactively presented, for example in stakeholder dialogues and publications such as annual and sustainability reports (Anglo Platinum Limited, 2006; PPC, 2006; Sasol, 2006; Exxaro, 2008; Anglo Platinum Limited, 2010; Arcelor Mittal, 2011; PPC, 2011; Exxaro, 2012). However, in other cases, the EEA is not even mentioned in reports (AngloGold Ashanti, 2006; BHP Billiton, 2006; Implats, 2006; Mondi Group, 2007; Mondi Group, 2008; BHP Billiton, 2010; AngloGold Ashanti, 2011; Implats, 2011; Mondi Group, 2011; Sappi, 2011; Sasol, 2011). This contradicts the reputational usage of the accord.

Non-members do not acknowledge the reputational effects of the EEA. On the one hand, the potential of the EEA in this respect is not recognised. This is also true for other activities with climate implications where some non-members have shown a rather late start and have not been seeking public attention (Non-Member, 1 2012). On the other hand, non-members deny that the South African public and customers have a positive attitude towards initiatives that focus on climate change mitigation or environmental protection (Non-Member 2, 2012). A comparably smaller degree of socialisation of the norm is therefore perceived by the non-members.

Altogether, reputational gains can be considered a crucial driver for EEA membership. The respective norm appears to be sufficiently socialised to be action-guiding for businesses.

Culture and leadership commitment

The analysed businesses are very aware of the climate change discourse. This is true for EEA members and non-members. All reject any notion of scepticism concerning the risks of climate change and its anthropogenic cause, and all already feel the need to adapt. They also share the opinion that 'something' has to be done against climate change, and all show a willingness to contribute (Mondi Group, 2007; NBI, 2007; Mondi Group, 2008; Implats, 2011; Mondi Group, 2011; NBI, 2011; Expert 1, 2012; Member 3, 2012; Member 4, 2012; Non-Member 1, 2012).

The connection between energy efficiency measures and climate change mitigation is also acknowledged. Annual and sustainability reports integrate the description of EEA membership into chapters on climate change and related issues (Exxaro, 2008; Anglo Platinum Limited, 2010; Arcelor Mittal, 2011; Mondi Group, 2011; NBI, 2011; Exxaro, 2012). In addition, some reports that do not include the EEA, or a direct connection between EEA and climate change, do include a connection between energy efficiency and climate change mitigation that is, however, also true for some non-members (AngloGold Ashanti, 2006; Arcelor Mittal, 2006; Sasol, 2006; AngloGold Ashanti, 2011; Sappi, 2011; Sasol, 2011). Likewise, companies that are not members of the EEA see potential on the part of business to contribute to mitigation and show willingness to do so (Richard Bay Minerals, 2006; Evraz Highveld, 2010; Gold Fields Limited, 2010; Xstrata South Africa, 2010; Harmony Gold Mine Company, Ltd. 2011; Non-Member 1, 2012; Non-Member 2, 2012). Yet, climate change is not considered a priority in the face of other more pressing problems and the perception that South Africa produces a comparably small share of global emissions. Here, it is emphasised that mitigation activities might have adverse effects on needed growth, economic development and poverty eradication (NBI, 2011; Member 3, 2012; Member 4, 2012; Non-Member 1, 2012; Non-Member 2, 2012). Therefore, other norms prevail. Also, climate change awareness of EEA members has been increasing, and was less pronounced in 2005 when the EEA was introduced (Implats, 2006; Sappi, 2006; Mondi Group, 2007; Mondi Group, 2008; Implats, 2011; Mondi Group, 2011; Sappi, 2011; Member 4, 2012). Therefore, the effect of a normative driver on motivating original participation in the EEA must be considered limited.

Yet, the EEA is also considered to be positively affecting the climate change discourse. It 'makes people think about the issue' which is named as a reason why the EEA is supported (Member 4, 2012). Non-members do not expect positive impacts of collective endeavours or awareness-raising through agreements such as the EEA (Non-Member 1, 2012; Non-Member 2, 2012). They also partly appear to have started engaging with the issue and possible own-activities later than members; for example non-members are only starting to develop strategies that include references to energy efficiency in the context of the eco-system and of climate change and they pay less attention to the issue in their annual reports (AECI, 2006; Evraz Highveld, 2006; Gold Fields Limited,

2006; Rand Water, 2006; Richard Bay Minerals, 2006; Transnet, 2006; AECI, 2010; Evraz Highveld, 2010; Gold Fields Limited, 2010; Richard Bay Minerals, 2010; Rand Water, 2011; Transnet, 2011; Non-Member 1, 2012). In addition, the fear of losing competitiveness, particularly regarding emerging economic powers in Asia, is very pronounced. Therefore self-restricting agreements, as well as government provisions, are not supported unless competitiveness aspects are sufficiently respected (Non-Member 1, 2012; Non-Member 2, 2012).

All together, climate change mitigation as an internalised norm is a rather weak, and certainly not the primary, driver for collective action in the EEA framework.

Norm internalisation and competitiveness considerations

Competitiveness considerations in the face of own commitments to energy efficiency measures for climate change mitigation cannot be considered a driver for EEA participation. The EEA members did not mention the hope that their direct competitors follow the same rules and make the same concessions as they do due to their membership in the EEA. Their pursuit of a collective agreement is therefore not aimed at preventing free riding.

Home country regulations

'Group-wide' or international strategies and commitments, which provide policies for issue areas such as energy consumption or environmental impact, influence practices and reporting (Mondi Group, 2008). Furthermore, companies standardise their environmental policies worldwide, and apply guidelines such as EU directives to their operations in South Africa (Member 2, 2011). Although the EEA membership may be in line with an international policy, the company representatives reported that this was not connected to direct mandates from headquarters in other countries. Also, home country regulations might lead to certain energy efficiency measures but there is no intention to foster the EEA as a collective agreement in order to secure competitiveness.

Mimicry

Mimicry is generally not found to be a driver for EEA participation of energy-intensive users. This is interesting because there are only slight differences in the level of other CSR commitments between members and non-members of the EEA (Expert 1, 2012). While companies might follow their peers regarding other measures, this seems not to be true for the EEA. Some non-members are not even aware of other companies'

EEA commitment. It is believed that almost 'nobody has signed it' and that it is actually not 'implemented' (Non-Member 1, 2012; Non-Member 2, 2012). It is therefore possible that they would have seen a stronger incentive to join the EEA if they had known about the number of signatories. However, it can be assumed that mimicry has not been a crucial driver.

Knowledge and learning

Information sharing, collaboration and learning are frequently mentioned as drivers for EEA membership. This function is considered particularly helpful in sectors and in relation to the government where distrust often impedes collaboration outside of institutionalised contexts. The awareness of the same risks and problems, especially regarding energy security and climate change consequences, has led to the acknowledgement that cooperation could benefit all affected. Thus, while outside the agreement information sharing is rare, a 'culture' of knowledge sharing and even a certain 'belonging' have evolved within the EEA. Workshops and other meetings are used as opportunities to work on common concerns (NBI, 2011; Expert 1, 2012; Member 3, 2012; Member 4, 2012). However, other collective arrangements are also considered equally apt for such cooperation (Member 3, 2012; Member 4, 2012).

Non-members who were questioned generally tend to prefer individual action (Non-Member 1, 2012; Non-Member 2, 2012). Not all entirely neglect the possible benefits of collaboration (Evraz Highveld, 2010; Non-Member 1, 2012), but the EEA is not considered an appropriate vehicle for mutual benefits due to specific operational contexts. Individual measures that are geared to the specific needs are considered more useful (Non-Member 1, 2012; Non-Member 2, 2012).

Thus, if companies believe that certain concerns and problems are shared, an institutionalised framework, such as the EEA, can be seen as a chance to exchange information for mutual benefit irrespective of a general level of mistrust. Hence, information sharing and learning can be considered as a driver for EEA membership.

Conclusions

In the sections above we analysed why companies voluntarily and collectively join the EEA. The framework of analysis is based on eight drivers that could foster cooperation on energy efficiency among South African Companies. Importantly, these drivers are not treated as distinct alternatives. Rather, several potential inter-linkages and mutual influences are taken into account. The drivers are subsequently explored by using a variety of sources: expert interviews, interviews with company and business association members, formalised interviews and analysis of the company's sustainability and annual reports. The findings are summarised in Table 5.1.

As Table 5.1 shows, there are two primary drivers for EEA participation: reputation and information sharing. These two aspects are the only ones for which substantive variation between members and non-members can be recognised. In respect of reputational gains by the EEA, all companies studied assume that a norm of climate-friendly behaviour is sufficiently socialised as to guide consumer decisions. Therefore, the EEA is used to communicate such behaviour and to benefit from reputational gains. Thus, irrespective of the individual internalisation of such a norm, its action-guiding power drives EEA participation to a substantial extent. Another important aspect that is frequently found as a motivation for EEA participation is the potential of such a collective agreement regarding knowledge sharing and collaboration in the face of generally distrustful relations between different companies as well as between companies and the government. The institutionalised framework of the EEA offers a vehicle for collaboration that is perceived as providing mutual benefits regarding shared challenges and risks such as climate change consequences, energy insecurity and increasing electricity price pressure.

Among the different drivers, energy security is most often mentioned by the interviewees and the publications reviewed. This is not surprising, as the electricity blackouts in recent years put South African industry in a very uncomfortable position. While energy efficiency issues in terms of costs could have been handled by each company themselves,

Table 5.1 Comparison of motivations for members and non-members

Immediate drivers	Members	Non-members
Shadow of anarchy	++	++
Shadow of hierarchy	+	+
Culture and leadership commitment	+	+
Norm internalisation and competitiveness considerations	–	–
Reputation	++	–
Home country regulations	–	–
Mimicry	–	–
Knowledge and learning	++	–

the energy shortages required a common approach. But while only collective action can stabilise the energy supply, the need for energy security does not differentiate between members and non-members of the EEA. It cannot be the sole explanation for why some companies join the EEA while others do not.

The same is true for the pre-emption of future regulation and the internalisation of a climate protection norm. When it comes to individual internalisation, the recognition of climate change as a relevant problem and the internalisation of norms differ only marginally between companies that participate in the EEA and those that do not. The same is true for the pre-emption of state regulation by participation in a voluntary agreement. The expected role of potential governmental sanctions is relatively low, but no difference was found in this respect between members and non-members.

Interestingly, both the 'shadow of anarchy' and the 'shadow of hierarchy' seem to affect business behaviour simultaneously. On the one hand, the state lacks the ability to provide the collective good of secure energy supply. On the other hand, the state is perceived as capable of introducing and enforcing laws that could provoke additional costs through taxes or energy cuts. Both 'shadows' can therefore coexist and motivate similar actions. Thus, complex issue areas like energy policy are related to different aspects of state capacity, which affects business behaviour through different channels.

Although all members state that direct cost-reduction through energy efficiency innovations is a main driver for EEA participation this cannot explain why a collective commitment is chosen over individual action. Although it is now clear that information sharing and reputational gains are the main drivers for participation in this collective agreement, it still remains to be established why some companies react to these drivers while others do not. While all interviewed companies agree on the need to adapt to and mitigate climate change, and all of them took some measures to increase energy efficiency, only some of them participate in the EEA. This leaves room for further research, especially on organisational drivers. Investigating the mechanisms that influence the receptiveness of institutional drivers would be the next research task.

References

AECI (2006) Annual Report 2006: Specialty product and service solutions, Sandton.
AECI (2010) Annual Report 2010, Sandton.
Anglo Platinum Limited (2006) Sustainable Development Report, Johannesburg.

Anglo Platinum Limited (2010) Sustainable Development Report, Johannesburg.

AngloGold Ashanti (2006) Report to Society 2006, Johannesburg.

AngloGold Ashanti (2011) Sustainability Report 2011: Sustainable Gold, Johannesburg.

Arcelor Mittal (2006) Sustainability Report 2006.

Arcelor Mittal (2011) Annual Report 2010: Bold Spirit.

Bansal, P. (2005) 'Evolving Sustainability: A Longitudinal Study of Corporate Sustainable Development', *Strategic Management Journal* 26, 3, 197–218.

BHP Billiton (2006) Sustainability Report.

BHP Billiton (2010) Our future: Sustainability Report.

Blanton, R.G. and Blanton, S.L. (2007) 'Human Rights and Trade: Beyond the "Spotlight"', *International Interactions* 33, 2, 97–117.

Börzel, T.A. and Risse, T. (2010) 'Governance without a State – Can it Work?', *Regulation and Governance* 4, 2, 1–22.

Chayes, A., Chayes, A.H. and Mitchell, R. (2000) 'Managing Compliance: A Comparative Perspective', in E.B. Weiss and H.K. Jacobson (eds) *Engaging Countries: Strengthening the Compliance with International Environmental Accords* (Cambridge, MA, MIT Press), pp. 39–62.

Di Maggio, P. and Powell, W.W. (1991) 'The Iron Cage Revisited: Institutional Isomorphism and Collective Rationality in Organizational Fields', in P. Di Maggio and W.W. Powell (eds) *The New Institutionalism in Organizational Analysis* (Chicago, IL: University of Chicago Press), pp. 63–82.

DME (2005) Energy Efficiency Strategy of the Republic of South Africa.

EIUG (2012) Energy Intensive User Group of Southern Africa. Retrieved 23.01.2012, from: http://www.eiug.org.za/.

Evraz Highveld (2006) Annual Report 2006. Mpumalanga.

Evraz Highveld (2010) Annual report for the year ended December 2010. Mpumalanga.

Expert 1 (2012).

Expert 2 (2012).

Expert 3 (2012).

Exxaro (2008) Exxaro Annual Report 2007. Pretoria.

Exxaro (2012) Annual Review 2011. Pretoria.

Flohr, A., Rieth, L., et al. (2010) *The Role of Business in Global Governance. Corporations as Norm-Entrepreneurs* (Basingstoke: Palgrave).

Gold Fields Limited (2006) Sustainable Development. Annual Report: 51–104.

Gold Fields Limited (2010) Sustainability Report. Annual Report: 91–125.

Hall, P.A. and Soskice, D. (2001) *Varieties of Capitalism. The Institutional Foundations of Comparative Advantage* (Oxford, Oxford University Press).

Hamann, R. and Börzel, T.A. (2013) (Chapter 1 in this volume). 'Introduction', in R. Hamann and T.A. Börzel (eds) *Business Contributions to Climate Change Governance in Areas of Limited Statehood* (London: Palgrave).

Harmony Gold Mine Company Ltd (2011) Sustainable Development Report 2011. Randfontein.

Haufler, V. (2001) *A Public Role for the Private Sector – Industry Self-Regulation in a Global Economy* (Washington D.C.: Carnegie Endowment for International Peace).

Héritier, A. and Lehmkuhl, D. (eds) (2008) 'The Shadow of Hierarchy and New Modes of Governance', *Journal of Public Policy*, Special Issue, 28, 1.

Hönke, J., Kranz, N., et al. (2008) *Fostering Environmental Regulation? Corporate Social Responsibility in Countries with Weak Regulatory Capacity. The Case of South Africa* (Berlin: Freie Universität Berlin).

Implats (2006) Corporate Responsibility Report 2006.

Implats (2011) Sustainable Development Report 2011. Illovo.

Jones, M.T. (1999) 'The Institutional Determinants of Social Responsibility', *Journal of Business Ethics* 20, 163–179.

Mail and Guardian (2011) Business and government pledge commitment to energy efficiency.

Mayntz, R. and Scharpf, F.W. (1995) 'Steuerung und Selbstorganisation in staatsnahen Sektoren', in R. Mayntz and F.W. Scharpf (eds) *Gesellschaftliche Selbstregulierung und politische Steuerung* (Frankfurt; New York, Campus), pp. 9–38.

Member 2 (2011).

Member 3 (2012).

Member 4 (2012).

Mondi Group (2007) Sustainability: Progress Report 2005–2006.

Mondi Group (2008) Mondi Group Sustainability Report 2007.

Mondi Group (2011) Mondi Group: Sustainable development review 2010 – Taking stock of our sustainability footprint.

Murphy, D. (2000) *The Structure of Regulatory Competition. Corporations and Public Policies in a Global Economy* (Oxford: Oxford University Press).

NEI (2007) Improved Energy and Efficiency at Mondi Business Paper – Richards Bay. Energy Efficiency Case Studies. Johannesburg.

NEI (2011) Case Studies: Anglo American. Energy Efficiency Case Studies 2011, Johannesburg.

NEI (2011) Case Studies: Anglogold Ashanti. Energy Efficiency Case Studies 2011, Johannesburg.

NEI (2011) Case Studies: BHP Billiton. Energy Efficiency Case Studies 2011, Johannesburg.

NEI (2011) Case Studies: Sasol. Energy Efficiency Case Studies 2011, Johannesburg.

NBI (2011) Energy Efficiency From: http://www.nbi.org.za/Focus%20Area/ClimateAndEnergy/EnergyEfficiency/Pages/default.aspx.

NBI and DME (2005) Energy Efficiency Accord.

Newell, P.J. (2001) 'Managing Multinationals. The Governance of Investment for the Environment', *Journal of International Development* 13, 7, 907–919.

Non-Member 1 (2012).

Non-Member 2 (2012).

Porter, M.E. and M.R. Kramer (2006) 'Strategy and Society: The link between Competitive Advantage and Corporate Social Responsibility', *Harvard Business Review* December, 1–15.

PPC (2006) Annual Report 2006: Embracing growth and transformation.

PPC (2011) Focused: Integrated Annual Report 2011

Prakash, A. and Potoski, M. (2006) *The Voluntary Environmentalists: Green Clubs, ISO 14001 and Voluntary Environmental Regulations* (Cambridge, Cambridge University Press).

Prakash, A. and Potoski, M. (2007) 'Investing Up: FDI and the Cross-Country Diffusion of ISO 14001 Management Systems', *International Studies Quarterly* 51, 3, 723–744.

Rand Water (2006) Annual Report 2006: Going Beyond Gold, Glenvista.

Rand Water (2011) Sustainability Report. Annual Report 2011, Glenvista.

Results from Formalised Interviews (2012).

Richard Bay Minerals (2006) 2006 Sustainable Development Report, Richard Bay.

Richard Bay Minerals (2010) Sustainable Development Report, Richard Bay.

Sappi (2006) Annual Report, Johannesburg.

Sappi (2011) Southern Africa Sustainability Report 2011, Johannesburg.

Sasol (2006) Technology for Sustainable Energy: Sasol Sustainable Development Report 2006, Johannesburg.

Sasol (2011) Pursuing Responsible Growth: Sustainable Development Report 30 June 2011, Johannesburg.

Smith, C.N. (2008) 'Consumers as Drivers of Corporate Social Responsibility', in A. Crane, A. McWilliams, D. Matten, J. Moon and D.S. Siegel (eds) *The Oxford Handbook of Corporate Social Responsibility* (Oxford, Oxford University Press), pp. 303–323.

Spar, D.L. and L.T. LaMure (2003) 'The Power of Activism: Assessing the Impact of NGOs on Global Business', *California Management Review* 45, 78–101.

Survey (2011) Results of a Company Survey Regarding the EEA Membership among EIUG: Conducted by the Authors (Berlin: Freie Universität Berlin).

Transnet (2006) Delivering on our commitments: Annual Report 2006 (including Sustainability Report) Johannesburg.

Transnet (2011) Sustainable Development Report, Johannesburg.

WBCSD (2008) National Business Initiative. Energy Effciency Accord. Collaborative Action for Energy: Regional Network Case Study.

Winkler, H., Davidson, O., Kenny, A., Prasad, G., Nkomo, J., Sparks, D., Howells, M., Alfstad, T., Mwakasonda, S., Cowan, B. and Visagie, E. (2006) 'Energy Policies for Sustainable Development in South Africa: Options for the Future'. Energy Research Centre, University of Cape Town, Cape Town.

Xstrata South Africa (2010) Sustainability Report, Johannesburg.

6
Business Contributions to Climate Change Adaptation – From Coping to Transformation? Insights from South Africa and Germany

Nicole Kranz

Introduction

So far, most research concerning the role of business vis-à-vis climate change has been targeting the issue of mitigation, that is the reduction of climate-relevant emissions and related strategic considerations (Kolk and Pinkse, 2008; Eberlein and Matten, 2009; Pinkse and Kolk, 2010). This chapter aims to take a different perspective, by focusing on the strategies and measures that assist with adapting to climate-induced environmental changes. It adds to literature on climate change adaptation from a governance (Dietz, Ostrom et al., 2003), as well as a business management perspective (Berkhout, Hertin et al., 2006; Winn, Kirchgeorg et al., 2010).

The chapter compares firms' strategies with regard to climate change adaptation in two very different settings: South Africa and Germany. South Africa is an emerging economy, with a quite well developed legislative framework with regard to water and environmental issues, but handicapped by weak capacities in terms of implementing legislation and by weaknesses in terms of sanctioning firms' malpractice. For this reason, South Africa is considered an area of limited statehood (Risse and Lehmkuhl, 2005). Germany is a country with consolidated statehood, with modern environmental legislation (for the most part aligned with EU legislation), as well as a relatively good track record in terms of effecting firms' compliance with environment and climate change related legislation. Eberlein and Matten (2009) show, for the field of climate change mitigation, that Germany has reached a high level of 'sophistication' in terms of the regulative measures issued, as well as compliance among firms with these regulations.

We might thus expect to find firms in Germany showing more advanced and far-reaching reactions and strategies to climate change in terms of adaptation, while in South Africa, we might expect to detect some activity in terms of adaptive action, albeit to a more limited degree. The findings of this research, however, point to different mechanisms. South African firms prove to engage in more numerous as well as more proactive activities targeted at climate change adaptation than do German firms. The chapter will explore the underlying reasons for this difference. Particular attention will be paid to the effects of a shadow of hierarchy or shadow of anarchy, as well as the role of problem pressure prompting firms to respond.

Accordingly, this chapter addresses the following research questions:

1. What are manifestations of business' adaptation efforts in different governance contexts?
2. What drivers or motivations lead to certain business adaptation efforts?

The research leads to a better understanding of business contributions to climate change adaptation in different governance contexts, ranging from mere coping measures to more transformative approaches (Hamann, Börzel et al., 2012). It places specific emphasis on the exchanges and interactions between firms and state, as well as other non-state actors, thus covering not only firms' internal adaptation measures, but also those company-driven activities that have implications for surrounding communities and are targeted at improving overall adaptive capacity.

The chapter focuses specifically on issues related to climate change adaptation in the water sector. Water resources management is considered to play a pivotal role in adaptation, as water resources and related (public and ecosystem) services are likely to be affected by climate-related impacts to a substantial degree (UN Water, 2010). Water-related impacts of climate change occur due to the changes in water quantity and quality, which arise from changes in the frequency, intensity and distribution of precipitation, as well as from evaporation rates. Longer dry periods are likely to reduce groundwater recharge and minimum flows in rivers, and thus affect water availability for agriculture, drinking water supply, manufacturing and energy production (Hoff, 2011). In most countries, water resources are subject to complex governance systems, which are often traditionally dominated by the state (Rogers and Hall, 2003; Tropp, 2007). Managing water resources under climate change conditions will require more adaptive management approaches that include non-state actors. This chapter focuses on the role of business in fostering the adaptive capacity of water management systems in South Africa and Germany.

In order to develop the two case studies, interviews were conducted during a two-week interview phase in South Africa in September 2011. These research results were compared and contrasted with initial research conducted in South Africa in 2008. Information about the German situation was collected through expert interviews and literature research in January 2012. Interviewees represented businesses and government at different levels as well as non-governmental stakeholders and independent experts. In terms of the industry sectors investigated, the aim was to look at a range of non-water businesses (i.e. no water utilities), which however have a considerable impact on water resources and display a certain dependency on the resource, and thus also have a potential to influence water policy. Sectors investigated included mining, food and beverage, general manufacturing and energy utilities.

Structuring firm contributions and drivers

The chapter makes use of categories and concepts developed in Chapter 1 of this book. Accordingly, firms' activities regarding climate change adaptation are allocated to four distinct categories, with a specific focus on water management.

1. Coping is targeted at increasing the short-term resilience of companies to climate change, that is, measures at the firm, in its supply chain or at the community level that help better manage scarce water resources (Folke, Carpenter et al., 2002). This may be, for example, through monitoring ecological or water systems (Dietz, Ostrom et al., 2003) or application of measures for increased water-efficiency within existing processes.
2. More substantial adjustments help increase resilience in the long term, for example through the development of innovative processes and technologies that allow for the reuse of wastewater and water savings. Water resources planning activities reaching beyond the immediate sphere of influence of a firm, include modelling and scenario-building. Firms become involved in activities that relate to their supply chain and the wider community by supporting large-scale monitoring efforts as well as planning processes through their expertise.

   Other measures, with substantial impacts beyond the company fence, refer to the development of water infrastructure, which aids in transporting and storing water, thus balancing potential impacts of climate change.

3. Transformation comprises more systems-orientated approaches to increasing the resilience and preparedness of firms and communities alike, through facilitating a more broad-based transformation. For these, firms might get involved in shaping political processes, targeted at supporting overall adaptive strategies of communities, including contributions to financing adaptive measures and the strengthening of those institutions managing transition processes (Ostrom, 2005), but also involvement in education and awareness-raising activities.

Turning to potential drivers for firm behaviour, as discussed in Chapter 1, I draw on drivers introduced by the management literature emphasising the relevance of efficiency gains due to proactive firm behaviour with regards to environmental issues (Porter and Linde, 1995; Porter and Kramer, 2002), but also highlighting the role of institutional factors (Hoffman, 1997). The difference in the strength and rigour of regulation between Germany and South Africa allows a comparison of the relative influence of the shadow of hierarchy and the shadow of anarchy in the two countries (Börzel, 2007). Eberlein and Matten (2009) show that regulatory strength matters in determining firm behaviour in the case of climate change mitigation. Strict, coercive regulation has driven companies, for example in Germany, to comply with climate protection laws, and has triggered innovations, while weak regulation and lax enforcement in other countries has resulted in slack response on the corporate side, as there was no incentive for firms to secure competitive advantages through compliance. The ability or inability of the state to provide for appropriate institutional contexts not only determines whether firms become active in the first place, but also which actors they to cooperate with in addressing regulatory gaps.

While this concept is to some extent transferable to climate change adaptation policies and firm behaviour, two limitations apply. Adaptation is more difficult to regulate, as it affects many different sectors. In addition, firms are faced with multiple applicable regulations, which might be related to and relevant for climate change adaptation. This means that it is difficult to relate firm strategies to one specific regulatory driver. For this reason, other motivations often weigh in more significantly.

Under conditions of weak regulatory capacity, firms that are dependent on scarce common-pool resources, often become active if their access to common-pool resources is threatened. A firm's actual or potential exposure to the impacts of climate change will significantly determine the intensity of involvement with activities targeted at improving adaptive capacity (Berkhout, Hertin et al., 2004). Firms might decide to collaborate with government as well as other stakeholders in order to manage specifically complex tasks.

Other institutional factors also play a role in driving firm behaviour with regards to adaptation. These include international norms, for example those generated through the United Nations Framework Convention on Climate Change (UNFCCC) negotiations,[1] which mingle with reputational concerns. Pressure also derives from local neighbouring communities, which are equally affected by climate change phenomena, and which directly compete with business for scarce resources (Pegram, Orr et al., 2009). In many cases, firms cannot ignore community interests, as climate change often affects a range of different actors, thus creating interdependences.

A further decisive factor is the comparative capability vested in firms for a) identifying problems, b) adapting internal strategies and c) for availing respective resources to implement solutions (Sharma and Vredenburg, 1998). Methner (Chapter 7) discusses the learning capacity of corporate actors, which allows for the continual adjustment of corporate strategies in response to climate change. This ability, however, needs to be complemented by appropriate resources actually to realise adaptive practices.

In the following sections, I will briefly outline the current policy environment regarding climate change and, in particular, adaptation in South Africa and Germany, and then proceed to map firms' contributions to climate change adaptation in these two countries.

Climate change and water: Administrative frameworks in South Africa and Germany

South Africa and Germany are faced with considerably different challenges as regards climate change adaptation, and they display fundamentally different approaches when dealing with climate change related impacts. The different coping strategies of the two countries reflect not only the different degrees of exposure and severity of expected impacts, but also the two countries' statehood: consolidated statehood with mature administrative systems in Germany versus an emerging economy with limitations in terms of implementation capacity in South Africa.

South Africa

Climate change is projected to have significant impacts on water resources management in South Africa (Mukheibir, 2008; Midgley, Scholes et al., 2011). According to the 2005 Environmental Outlook for South Africa

(DEAT, 2005), climate change is expected to lead to significant alterations of current hydrological regimes and thus limit the capacity to adapt to further climate-related and other impacts (Mukheibir, 2008). Due to the high variability of rainfall and evaporation data, it is difficult to identify clear trends; and this leads to a high degree of uncertainty in predictions regarding the behaviour of water systems under increased climate stress (Mukheibir, 2008). An example of this uncertainty is the divergence of rainfall predictions. Intergovernmental Panel on Climate Change (IPCC) projections (IPCC, 2007) predict a further decrease in rainfall for the eastern part of the country – more down-scaled models predict an overall drying trend for the south-western part of the country, while, particularly in summer and autumn, northern and eastern parts might become wetter. There is, however, agreement that the overall trend, featuring higher temperatures of 1–3 °C, and higher evapotranspiration, points to a net reduction in available water resources leading to negative impacts on all water users (Midgley, Chapman et al., 2007).

According to projections (Addams, Boccaletti et al., 2009), South Africa's economic development could be severely constrained by limited water resources due to climate change impacts in the near future. Using a lower growth perspective than that proposed by Asgi-SA,[2] water for all current uses can only be assured with a high reliability until 2010. The scenario of the 2030 Water Resources Group foresees a gap between projected demand and supply of 17 per cent across all sectors.

Agriculture is one of the main water users (about 62 per cent of overall water demand), followed by urban/municipal water uses (23 per cent); power generation, mining and bulk industry account for about 8 per cent of total water demand. With a total water demand of 13 to 14 billion m³ the totally available amount has almost been used to the maximum (Addams, Boccaletti et al., 2009). Water demand for household use is projected to increase significantly, with larger shares of the population attaining access to water services. Industrial water demand is likely to double over the next 20 years. Specifically, meeting water demand for power generation constitutes a challenge, as water-intensive coal combustion remains the main source of energy for the years to come. Agriculture will continue to constitute an important element of the economy, thus necessitating the development of more water-efficient irrigation methods in order to enable increasing productivity to ensure food security. The mining sector's water requirements are also expected to grow over the next 20 years for production purposes, as well as for diluting acid mine drainage (Addams, Boccaletti et al., 2009). These aspects also point to additional stresses imposed on South Africa's water

resources due to mounting pollution originating from industrial effluents, domestic and commercial sewage, acid mine drainage and agricultural runoff (DEAT, 2008).

Accordingly, the overall policy setting regarding climate change is
changing rapidly. The South African government has prepared a Green
Paper on climate change adaptation[3] and is currently engaged in a consultation process leading up to a White Paper, which will then inform actual
policies and follow-up legislation. In the run up to the 2011 Durban COP,
the South African government engaged in several stakeholder processes,
also involving business actors, to develop a distinctly South African perspective of climate change adaptation. The lead department for climate
change adaptation is the Department of Water Affairs, where a Director
for Climate Change Issues has been appointed. Efforts related to climate
change adaptation need to be coordinated with existing legislation on
water management. South African water policy is guided by the principles of integrated water resources management and thus features quite
advanced provisions for climate change adaptation. Nevertheless, the
implementation of these policies has posed a major challenge, as in
many cases the provisions turned out to be too demanding. In addition,
government departments charged with implementation, particularly at
the municipal level, struggle continuously due to limited staff and funding (Kranz, 2010). While implementation records are slowly improving,
the increasingly demanding requirements are likely to put additional
pressure on government departments, especially since it requires far-
reaching cooperation among different portfolios.

All these factors create a considerable level of uncertainty for companies operating in South Africa, either because of a serious threat to their
resource-base, due to climate variability, or because of unforeseeable
development in the regulatory space.

Germany

Climate change impacts are expected to become noticeable in Germany
as temperatures increase by 1.5 and 3.7 degrees. More heat waves are
expected, together with a decline in summer precipitation by about 30 per
cent (UBA, 2008). Climate change impacts are furthermore expected to
vary significantly across regions. For example, the north-eastern regions
of Germany, in particular, which are already affected by droughts, are
projected to experience even less summer rainfall. Projections for other
regions predict similar patterns, with less drastic effects, more attenua-
tion and less certainty of prediction. Expected impacts in the health, agriculture, forestry, transportation and tourism sectors are, to a large extent,

related to the expected challenges arising for water management: either greater or reduced precipitation, greater risk of floods in winter and spring, more frequent low water in summer, and altered groundwater tables with possible consequences for water supply for domestic and industrial purposes (UBA, 2008).

In terms of the policy environment, at the EU level, the White Book on Climate Change Adaptation constitutes the overarching framework guiding adaptation policies at the national level in EU member states (EU, 2009).

The German federal parliament adopted the National Adaptation Plan, the so-called action plan 'Adaptation to Climate Change' in 2008 (UBA, 2008). This plan constitutes a strategic framework guiding all relevant actors in their concerted adaptation actions. The national adaptation strategy (NAS) adopts a comprehensive approach, ranging from the identification and communication of dangers and risks with regards to climate change adaptation, the creation of awareness to the provision of a basis for decision-making, enabling various actors to take precautions and gradually adapt to climate change impacts (Swart, Biesbroek et al., 2009). According to the NAS, six principles should guide action for climate change adaptation in Germany: openness and transparency, a flexible precautionary approach, subsidiarity and proportionality (regional and sectoral) integration, international responsibility (see below) and sustainability. While the NAS thus establishes a joint basis for action, it refrains from prescribing detailed adaptation scenarios and instead calls on individual responsibility and action by key actors. Business is identified as one of the key actors for carrying out some of the projected climate change adaptation issues. More specific actions are to be suggested by the planned National Adaptation Plan, which is to be published in due course. Overall, adaptation planning in Germany appears to range between specific ideas about measures to be taken, which are prescriptive in their overall direction, while at the same time leaving room for interpretation by different actors in their specific sphere of influence. Somewhat reflecting long-term political inaction on the adaptation part, the German government engaged in a rather slow process in drawing up the NAS. This hesitant approach is to some extent also reflected in the delayed implementation of the National Adaptation Plan (Swart, Biesbroek et al., 2009).

The Ministry for Environment is the lead department guiding national adaptation efforts. It maintains the so-called *Kompass*, Competence Center for Climate Change Adaptation, which serves as the main knowledge hub for adaptation-related information and research.[4]

Consequently, the overall frame for companies to become active regarding climate change is well defined. Even though climatic changes to be expected are uncertain, firms can rely on relatively well-developed information sources. The National Adaptation Strategy furthermore is designed to incentivise firms to become active.

Business and climate change adaptation in South Africa and Germany

The findings on company strategies in South Africa and Germany are grouped in the sub-categories for business behaviour vis-à-vis climate change adaptation as outlined in the introduction of this book.

From coping to transformation? – The case of South Africa

Coping

Many firm activities are focused on coping and mostly pertain to conventional water management measures targeted at improving the firms' ability to cope with climate change impacts.[5] Corporate risk reduction strategy involves, in part, activities to increase internal water use efficiency during production processes, including recovery and recycling of waste water as well as securing access rights to water sources. Nestlé cut down water usage at its Mossel Bay dairy plant in the Eastern Cape by 54% between October 2009 and May 2010 in reaction to a severe drought, which affected communities and industries alike.[6] Significant savings were attained from the recovery and re-use of condensate from evaporation processes. In addition, several immediate water-saving measures were introduced, such as shortening automated wash time and modifying hosepipe nozzles[7] to reduce water flow. Employees were also advised on water-saving measures at home.

All interviewees highlighted the challenge of gauging the uncertainty deriving from climatic changes and the corresponding question of when to implement measures to adjust to potential climate variations. Corporate managers very often advocated a pragmatic approach, which would only allow for measures taken in response to present risks, that is, to cope with immediate threats to operations. Firms appeared to be more hesitant to engage in long-term measures, the effects of which would be uncertain.[8] They used this time perspective as a tool to differentiate measures into 'low hanging fruit' that could safely been undertaken 'here and now' and those with a more uncertain risk perspective in the long-term.

Substantial adjustments

While coping measures were particularly high on the agenda of South African firms, there is, however, also evidence of activities that support more substantial adjustments in the longer term. Examples of this include contributions to catchment management functions, in collaboration and with the support of government,[9] such as the sharing of monitoring data – for example, on water flows. While SABMiller's activities are a prominent case in this regard, overall there is an emerging awareness of the potential role of firms in these contexts, pointing to the heightened interest of firms to become engaged with activities beyond their organisational boundaries.

Activities supporting substantial adjustments also occur along the firm's supply chain or its immediate operating environment. An example of a firm's interaction with communities is the Replenish Africa Initiative implemented by Coca-Cola in 100 villages across South Africa. This project is mainly aimed at improving access to water (and sanitation) in schools and other public buildings and has an obvious philanthropic component. While not affecting the company's own water management practices, it still assists communities in fostering their ability to adapt to climate change impacts through more stable water supply systems.

More far-reaching approaches include Nestlé's farmer programmes, which provide guidance to the company's suppliers on the efficient use of natural resources, including water management at the farm level. In this programme, farmers are educated about sustainable natural resource use and adequate monitoring and reporting. Similarly, SABMiller assists smallholder hops and barley farmers in their supply chain to develop skills that enable them to engage in sustainable water management practices.

The development of water infrastructure, for example storage reservoirs and dams, is a common way to react to and anticipate climatic variations and has been prominent in South Africa for some time (Turton, Schultz et al., 2006; Matthews, Wickel et al., 2011). Mining companies, in particular, have recently become involved with infrastructure development designed to increase storage capacity or to free up freshwater contingents for municipal use, thus improving overall water availability under conditions of water stress. Examples can be found in the platinum industry, where business actors helped to leverage financial resources for the construction of water infrastructure in the context of cross-sectoral partnerships between mining firms and national, provincial and local government (Kranz, 2010).

While such interventions clearly allow for substantial adjustments, creating a basis for increasing resilience to climate change impacts, it needs to be noted that infrastructure developments need to be accompanied by careful planning efforts in order to ensure the selection of the correct infrastructure option and assure long-term sustainability (Matthews, Wickel et al., 2011). As a consequence, firms also often become involved with planning activities with regards to water infrastructure development, cooperating closely with government representatives at different levels. Joint planning is necessary for moving beyond the immediate interests of a firm (e.g. the extension of bulk infrastructure for corporate water security) to meet community needs (e.g. infrastructure for small-scale water storage and reticulation). Supporting joint planning activities through know how and technological and managerial expertise can thus be considered a firm activity supporting substantial adjustments.[10]

Transformation

More transformative approaches, which are targeted at more systemic and long-term changes, are still emerging in South Africa. Some evidence can be found in the following examples. Sasol's engagement with local government to promote water-saving measures in the Emfuleni municipality (Gauteng Province) is an instructive example of a firm's role in promoting joint learning that contributes to increasing the resilience of water-stressed communities.[11]

The example of SABMiller, on the other hand, demonstrates that internal initiatives targeted at increasing a firm's resilience can lead to learning and transformative approaches for other actors as well. Based on detailed water footprint assessments for SABMiller operations and the relevant agricultural suppliers, a preliminary basin-wide risk analysis and sustainability assessment was conducted, taking account of key ecosystem services and water needs, as well as future threats from economic development and climate change. This formed the foundation for the engagement with other actors in a multi-stakeholder collaboration to coordinate the water-conservation initiatives of various parties, in order to ensure realisation of synergies in designated basins and develop resource-protection initiatives (Water Futures, 2011). In the context of this initiative, the firm contributed more than just an alignment of different stakeholders' interests. It worked at the international level to develop an innovative method of assessing firms' water impact; and in a second phase of the project, it promoted the widespread application of this methodology (Water Futures, 2011). This initiative is clearly exceptional in terms of its scope and foresight and thus marks the upper end of the spectrum of possible contributions.

Nevertheless, having businesses engage with public policies pertaining to climate change adaptation, in a way that increases the overall system's resilience, is currently gaining momentum in South Africa. This type of involvement includes business's role in river basin (or catchment) management, as well as interaction with government at different levels on specific climate change related topics, for example the context of advisory forums or other consultation processes.

The role of business in catchment management planning is still limited, although the necessity to involve water-intensive businesses is increasingly recognised by government agencies, NGOs and consultants. In the Western Cape Province, a project is underway to facilitate a government–business dialogue on the provincial economy's reliance on water, and on how best to respond to the resulting risks and opportunities.[12] As a start, business actors are represented in the governing board of the Breede catchment management agency,[13] one of the agriculturally most relevant catchments of the Western Cape. Firms involvement is however only evolving quite slowly. SABMiller's engagement in the alien vegetation clearing,[14] and its cooperation with the Caledon municipality on water-use efficiencies, stand out as a leading examples. Other beverage firms are expected to follow suit in the fear that 'they would feel the pain' of decreasing water resources.[15]

Comparatively more intensive is the engagement of some firms with both provincial and national government, on water and climate change related issues. Business representatives[16] were appointed as members of an advisory group for the water minister on mitigation and adaptation issues, set up to develop strategies for adapting to climate change across all uses and sectors. Business representatives are also involved in consultations regarding the Green and White Paper on Climate Change Adaptation (see above). Several water intensive businesses have recently joined forces in the Strategic Water Partners Network South Africa in order proactively engage with government as well as other partners on closing the anticipated water supply and demand gap through coordinated efforts (NEPAD Business Foundation, 2012).

Focus on coping: The German case study

General comments

There is limited documentation of the activities related to climate change adaptation by German companies. Some literature is now emerging in the context of regional adaptation studies. Overall, there is considerably less activity in Germany than South Africa, and Germany shows a strong focus on reactive approaches or coping strategies. We find very little evidence of approaches beyond coping. Mitigation features much more prominently on the agenda of German companies; see for example the initiative '2 Degrees' by German entrepreneurs, which is targeted at crafting a coordinated strategy for mitigation.[17]

In the following section, some examples of company strategies for adaptation will be discussed for the energy/utilities sector (Stecker, Pechan et al., 2011) and for manufacturing.

Coping

Initial coping strategies are quite common among companies in the utility sector. They include, for example, monitoring processes to gauge potential climate change impacts on operational factors such as water availability. Overall, however, companies are mostly at the stage of developing strategic approaches based on perceptions of climate change impacts.[18] In many cases, companies reported that they have been discussing the issue internally in order to gauge impacts and to prepare potential measures. In a few instances, these discussions have been followed by concrete planning efforts. Often, these have been aligned with, and added on to, existing planning and monitoring tools, such as the observation and prediction of extreme weather events. In this regard, larger firms displayed a quite advanced level of preparation, since they already keep track of the key influencing factors of their operations (Pechan, Rotter et al., 2011).

In terms of concrete measures assisting in actually coping with climate change impacts, firms in the utility sector have resorted to quite conventional approaches, for example improvement of forecast tools and reconsideration of infrastructure developments, such as the enlargement of water storage capacity, new strategies for cooling of thermal and nuclear power plants.[19]

In the manufacturing sector, current activities are also mostly focused on understanding impacts deriving from either extreme weather events or gradual climate change phenomena. Especially in the food, textiles and forest products sectors, there is an emerging debate on the need to improve adaptive behaviour, while only a few approaches have been implemented on the ground (Stechemesser and Günther, 2011). Firms are increasingly integrating climate change considerations in their environmental management systems.

Substantial adjustments

Contributions to substantial adjustments, such as those emerging in South Africa, have not been identified in the German context. Some firms, such as BASF, have started developing and applying innovative technologies and materials that help prevent flooding (material solutions for enhanced floodwalls) and breed stress-tolerant plants capable of coping with extreme weather events. As these products are mostly sold in foreign markets, product development is in some instances combined with capacity-building and technology transfer.[20]

German food retailers are slowly developing measures promote more water-considerate strategies among their suppliers of agricultural products, especially within Germany and to some extent also for suppliers in other regions across the world. Firm activities to control and influence overall resource use along the supply chain can also be regarded as measures that potentially support adjustments to climate change. In many cases, however, activities are motivated by a basket of objectives; climate change objectives are included alongside those relating to food quality and safety (Beddington, Asaduzzaman et al., 2012).

In sectors, where firms are required by default to cooperate intensively with government actors, for example energy utilities, adaptation measures issued by these companies are likely to have far-reaching effects on other actors as well. Through such adaptation mainstreaming measures broad-based and substantial adjustments for a number of actors could be triggered through corporate behaviour (Pechan, Rotter et al., 2011).

Transformation

Activities that could lead to the systematic transformation of water management approaches, in anticipation of potential climate change impacts, are rare among German companies. In some instances, in response to extreme weather events (e.g. drought in the Rhine basin), utility companies have begun to interact more intensively with water managers.[21] There is, however, very limited engagement in policy development. Although the National Adaptation Strategy mandates the involvement of a broad range of stakeholders, corporate actors have been involved only to a limited extent: in several workshops convened by the Ministry for the Environment, as well as in some emergent discussions.

What drives these strategies?

Findings from case studies indicate that South African firms are more proactive than companies investigated in Germany. This finding is unexpected, given the high level of activity of German companies around the issue of climate change mitigation, as well as the fairly well-developed framework for action on climate change adaptation proposed by the German government over the past years. The South African business community, or at least some of the larger, eminent companies, appear to be proactive in tackling climate change adaptation and shaping policies that transcend their immediate sphere of influence and have a potential to transform overall policies. What drivers explain the variation between Germany and South Africa?

The shadow of hierarchy, or its absence (shadow of anarchy), is one of the first factors to be examined. The policy landscape in South Africa is still only emerging. This leaves room for companies to become active on their own accounts. Specifically, this behaviour could be motivated by the following factors.

In the first instance, the actual or perceived inability of government to address certain issues is of importance. Firms are affected by the absence of national and often local government action to address adaptation through properly implementing existing water management regulation, providing for the necessary upgrading of water infrastructure, and resolving potential allocation conflicts among different water users. This directly affects firms' resource base, which is a strong underlying driver, especially for companies, such as mining or beverage companies, with a direct reliance on water.[22] It also determines the way companies address this issue: in cases where companies are mostly securing their short-term access to resources, they resort to coping measures, in cases where they take on a longer-term perspective, and are thus faced with more complex tasks, they engage with other actors.

In addition, there is anticipation that government might soon issue additional water regulation, targeted at adaptation, which is likely to impact firms' access to water resources or entail requirements in order to obtain access. The parliamentary White Paper and other policy documents point in this direction. Consequently, firms are attempting to position themselves accordingly, presenting potential solutions and shaping the overall policy discourse. Thus, it could be argued that a certain shadow of hierarchy is 'looming'; it can be considered an emerging threat given the poor track record of implementation in the past.

The situation is different in Germany, where a climate change adaptation strategy has been issued by government after prior consultation with a broad range of stakeholders. Thus, firms either have already had the opportunity to present their position, or find themselves involved in a continued, albeit low-intensity, policy dialogue on these issues, coordinated by government. In general, firms take a hesitant and reactive approach to government policies, partly due to the regulatory history in terms of environmental issues: the German government has taken quite a strong regulatory stance on mitigation issues, keeping firms occupied with compliance in this area. At the same time, firms also shy away from too much involvement with government in order to avoid additional oversight and control on adaptation. Also, firms have experienced few problems regarding adaptation so far (see context section above). In fact, most companies have had little exposure to the detrimental effects of climate change. As a consequence, there is limited awareness, and measures are evolving as part of precautionary action towards anticipated extreme weather events, rather than in response to actual damages or shortcomings (Pechan, Rotter et al., 2011; Stecker, Pechan et al., 2011).

While securing access to potentially scarce resource does not constitute a strong driver in the German context, another competitive aspect is slowly emerging as important for some companies, who seek to position themselves as innovators on water-saving and related technologies that might open up competitive markets in Germany or elsewhere. In this sense, firms emulate behaviour detected by Eberlein and Matten (2009) with regards to mitigation. It also explains the rather individualistic approach by companies, who are rather interested in securing their own competitive advantage rather than cooperating with other actors, as seen in the South African context. The exposure of firms to climatic changes, have made cooperation with other actors necessary in South Africa, in order to address complex tasks such as managing the impact of climate change in entire catchment areas. In Germany, on the other hand, firms see skills and expertise in climate change adaptation as a competitive advantage and business opportunity and thus far refrain from cooperation.

In addition, firms are embedded in significantly different normative contexts in South Africa and Germany. South African business is highly sensitised about potential impacts of climate change specifically on water resources. This attention is heightened, through the international debate (e.g. the 2011 Durban COP (see above)), activities of the CEO Water Mandate.[23] Large South African firms with strong ties to the international business community have become especially prominent participants in international debates. But climate change and water are also high on the agenda both nationally and locally. A factor that emerged strongly, especially in interviews with mining representatives, was the responsibility towards neighbouring communities that would drive certain activities in the water management field. While this awareness has only emerged slowly and over a long period, it clearly manifested as one of the key components in the respective companies' incentive structure.[24]

Here, too, the picture is different in Germany, where mitigation strongly dominates the policy discourse and topics such as emission trading feature prominently on corporate radar screens.[25] One could argue that companies, affected by an extreme shadow of hierarchy for mitigation, and hesitant to take on additional burden related to adaptation, are not yet exposed to peer discourses and public opinion on adaptation (Stecker, Pechan et al., 2011). Business strategies are slowly changing in reaction to the international debate, in part fuelled among others by South African companies.

Conclusions

This chapter has shown that firms adopt different strategies of climate change adaptation in South Africa and Germany, ranging from coping measures to more substantial adjustments, in some South African cases. The research also indicates that South African firms are more proactive in addressing adaptation as a core element of their strategies than are German firms, which have so far taken a rather reactive approach.

Proactive behaviour in South African companies is in the first instance driven by actual or projected resource constraints, with specific emphasis on water resources. In order to reduce their dependency on scarce resources affected by climate change, companies engage in efficiency measures. More far-reaching initiatives involve other stakeholders, including government, in jointly addressing the complex tasks around water resources management. German companies, on the contrary, have not yet been exposed to severe climate change impacts and thus remained relatively inactive.

Considering the governance context, firms in the two cases respond differently to their respective regulatory environment. Companies in South Africa are taking a proactive stance because of their concern that the government will not be able to ensure reliable water access under climate change conditions. In Germany on the other hand, governance tradition, in the water and environment sectors, is largely premised on hierarchical steering and command and control by the state. Thus companies expect this state role with regard to adaptation as well and wait for it to be developed, rather than becoming engaged in proactive behaviour.

Notes

1. See IISD Report of the Global Business Day at the 17th COP in Durban, November 2011.

2. South Africa's Accelerated and Shared Growth Initiatives maps out potential areas of growth for the South African economy.
3. National Climate Change Response Green Paper, Government of the Republic of South Africa, 2010.
4. Accessible at www.bmu.de/klimaschutz/anpassung.
5. Interview with Risk Mitigation Officer, Impala Platinum, September 25, 2011, Johannesburg.
6. Interview with Sustainability Manager, Nestlé South Africa, November 17, 2010; Cape Town at UN Global Compact meeting.
7. Nozzles are mechanic devices to regulate the flow in pipes and tubes. Technological changes help in managing flow more effectively.
8. Interview with Risk Mitigation Officer, Impala Platinum, September 25, 2011, Johannesburg. Interview with Environmental Manager, Sasol, September 25, 2011, Johannesburg.
9. The term catchment refers to a river basin and describes the entire scale of the water management unit.
10. An example for this can be found in the South African Coal mining industry. See Kranz (2010).
11. Interview with Environmental Manager, Sasol, September 25, 2011, Johannesburg.
12. Interview with Director, Pegasys Consulting, September 21, 2011, Cape Town.
13. The Breede-Overberg catchment area is situated in the Western Cape Province and is one of the two catchment areas in South Africa with a fully established Catchment Management Agency. The area is water-stressed and dominated by agriculture with risks around environmental requirements, water quality deterioration, and allocation of water for high-value agriculture for processing and export, reallocation of water to emerging black farmers.
14. The removal of alien, highly water-intensive vegetation in upstream parts of rivers is expected to relieve water stress in downstream parts.
15. Interview with Director, Pegasys Consulting, September 21, 2011, Cape Town.
16. Interview with Environmental Manager, Sasol, September 25, 2011, Johannesburg.
17. See website: http://www.initiative2grad.de/.
18. Interview with CR manager at German utility, 5 January 2012.
19. Interview with CR manager at German utility, 5 January 2012.
20. See pledges at the UNFCCC website in the context of the http://unfccc.int/adaptation/nairobi_work_programme/partners_and_action_pledges/.
21. Interview with CR manager at German utility, 5 January 2012.
22. Interview with Risk Mitigation Officer, Impala Platinum, September 25, 2011, see Kranz (2010). Interview with Sustainability Manager, SABMiller, November 17, 2010; Cape Town at UN Global Compact meeting.
23. Interview with Sustainability Manager, Nestlé South Africa, November 17, 2010; Cape Town at UN Global Compact meeting.
24. Interview with Risk Mitigation Officer, Impala Platinum, September 25, 2011, see Kranz (2010).
25. This might change drastically however after the devastating 2013 flooding events that severely affected numerous businesses.

References

Addams, L., Boccaletti, G., et al. (2009) Charting our Water Future – Economic Frameworks to Inform Decision-Making. 2030 Water Resources Group, McKinsey and Company.

Feddington, J., Asaduzzaman, M., et al. (2012) 'Achieving Food Security in the Face of Climate Change: Final Report from the Commission on Sustainable Agriculture and Climate Change', *A. a. F. S. CGIAR Research Program on Climate Change* (Copenhagen, Denmark).

Berkhout, F., Hertin, J., et al. (2004) 'Business and Climate Change: Measuring and Enhancing Adaptive Capacity', Tyndall Centre Technical Report. T. A. project, Tyndall Centre for Climate Change Research 11.

Berkhout, F., Hertin, J., et al. (2006) 'Learning to Adapt: Organisational Adaptation to Climate Change Impacts', *Climatic Change* 78, 1, 135–156.

Börzel, T.A. (2007) 'Governance Ohne Den Schatten der Hierarchie: Ein Modernisierungstheoretischer Fehlschluss? Regieren ohne Staat?', in T. Risse and U. Lehmkuhl (eds), *Governance in Räumen begrenzter Staatlichkeit* (Baden-Baden: Nomos).

DEAT (2005) South Africa Environmental Outlook, Department of Environmental Affairs and Tourism.

DEAT (2008) People – Planet – Prosperity: A National Framework for Sustainable Development in South Africa, Department for Environmental Affairs and Tourism.

Dietz, T., Ostrom, E., et al. (2003) 'The Struggle to Govern the Commons', *Science* 302, 1907–1912.

Eberlein, B. and Matten, D. (2009) 'Business Responses to Climate Change Regulation in Canada and Germany: Lessons for MNCs from Emerging Economies', *Journal of Business Ethics* 86, 241–255.

EU (2009) Anpassung an den Klimawandel: ein europäischer Aktionsrahmen. E. Commission. Brussels.

Folke, C., Carpenter, S., et al. (2002) 'Resilience and Sustainable Development: Building Adaptive Capacity in a World of Transformations', *AMBIO*, 437–440.

Hamann, R., Börzel, T.A., et al. (2012) *Business Contributions to Climate Change Governance in Areas of Limited Statehood*, ISA Annual Convention, San Diego.

Hoff, H. (2011) 'Understanding the Nexus', Background Paper for the Bonn 2011 Conference: *The Water, Energy and Food Security Nexus* (Stockholm: Stockholm Environment Institute).

Hoffman, A. (1997) *From Heresy to Dogma: An Institutional History of Corporate Environmentalism* (San Francisco, CA: New Lexington Press).

IPCC (2007) *Summary for Policymakers: Climate Change 2007: Climate Change Impacts, Adaptation and Vulnerability* (Geneva: International Panel on Climate Change).

Kolk, A. and Pinkse, J. (2008) 'A Perspective on Multinational Enterprises and Climate Change: Learning from "An Inconvenient Truth"?', *Journal of International Business Studies* 39, 8, 1359–1378.

Kranz, N. (2010) What Does It Take? Engaging Business for Addressing the Water Challenge in South Africa. Political and Social Sciences (Berlin: Freie Universität).

Matthews, J.H., Wickel, B.A.J., et al. (2011) 'Converging Currents in Climate-Relevant Conservation: Water, Infrastructure, and Institutions', *PLOS Biology* 9, 9.

Midgley, G.F., Chapman, R., et al. (2007) 'Impacts, Vulnerability and Adaptation in Key South African Sectors', LMTS Input Report. Cape Town, Energy Research Centre. 5.

Midgley, G.F., Scholes, R.J., et al. (2011) 'Scoping of the approximate climate change adaptation costs in several key sectors for South Africa up to 2050', Pretoria, SANBI, CSIR, ASSET Research, University of Pretoria.

Mukheibir, P. (2008) 'Water Resources Management Strategies for Adaptation to Climate-Induced Impacts in South Africa', *Water Resources Management* 22, 1259–1276.

NEPAD Business Foundation (2012) Strategic Water Partners Network South Africa: Closing the Water Gap by 2030, Johannesburg.

Ostrom, E. (2005) *Understanding Institutional Diversity* (Princeton: Princeton University Press).

Pechan, A., Rotter, M., et al. (2011) Anpassung in der Versorgungswirtschaft: Empirische Befunde und Einflussfaktoren', in A. Karczmarzyk and R. Pfriem (eds) *Klimaanpassungsstrategien von Unternehmen* (Marburg: Metropolis).

Pegram, G., Orr, S., et al. (2009) *Investigating Shared Risk in Water: Corporate Engagement with the Public Policy Process* (Surrey: WWF-UK).

Pinkse, J. and Kolk, A. (2010) 'Challenges and Trade-offs in Corporate Innovation for Climate Change', *Business Strategy and the Environment* 19, 4, 261–272.

Porter, M.E. and Linde, C.v.d. (1995) 'Green and Competitive: Ending the Stalemate', *Harvard Business Review* (September-October), 120–134.

Porter, M.E. and Kramer, M.R. (2002) 'The Competitive Advantage of Corporate Philanthropy', *Harvard Business Review* 80, 12, 56–68.

Risse, T. and Lehmkuhl, U. (2005) Teilprojekt A1: Governance in Räumen begrenzter Staatlichkeit: Beiträge zur Theoriebildung. Geplanter Sonderforschungsbereich 700. Antrag auf Einrichtung und Finanzierung 2006–2009. Governance in Räumen begrenzter Staatlichkeit? F. U. Berlin (Berlin: Freie Universität Berlin), pp. 91–140.

Rogers, P. and Hall, A.W. (2003) *Effective Water Governance* (Stockholm, Global Water Partnership).

Sharma, S. and Vredenburg, H. (1998) 'Proactive Corporate Environmental Strategy and the Development of Competitively Valuable Organizational Capabilities', *Strategic Management Journal* 19, 8, 729–753.

Stechemesser, K. and Günther, E. (2011) 'Herausforderung Klimawandel – Auswertung einer Deutschlandweiten Befragung im Verarbeitenden Gewerbe', in A. Karczmarzyk and R. Pfriem (eds) *Klimaanpassungsstrategien von Unternehmen2* (Marburg: Metropolis).

Stecker, R., Pechan, A., et al. (2011) *Why are Utilities Relctant to Adapt to Climate Change?* (Oldenburg: Berlin, Chameleon Research Group).

Swart, R., Biesbroek, R., et al. (2009) Europe Adapts to Climate Change: Comparing National Adaptation Strategies (Helsinki: Partnership for European Environmental Research).

Tropp, H. (2007) 'Water Governance: Trends and Needs for New Capacity Development', *Water Policy* 9, Supplement 2, 19–30.

Turton, A., Schultz, C., et al. (2006) 'Gold, Scorched Earth and Water. The Hydropolitics of Johannesburg', *Water Resources Development* 22, 2, 313–335.

UBA (2008) *Adaptation is Necessary – Germany in the Midst of Climate Change* (Dessau: Umweltbundesamt).

UN Water (2010) Climate Change Adaptation: The Pivotal Role of Water. UN Water Policy Brief, FAO Water.

Water Futures (2011) Addressing Shared Water Challenges through Collective Action, London, Eschborn: SABMiller, WWF-UK and GIZ.

Winn, M., Kirchgeorg, M., et al. (2010) 'Impacts from Climate Change on Organizations: A Conceptual Foundation', *Business Strategy and the Environment*.

7
Adaptation to Climate Change: An Investigation into Woolworths' Water Management Measures

Nadine Methner

Introduction

This chapter investigates how and why business organisations contribute to efforts to adapting to climate change in countries where the state is not always able or willing to fulfil its traditional governance functions (i.e. the provision of public goods and enforcement of rules) and struggles to respond effectively to the uncertainties and complexity of climate change. Given current concerns about the socio-economic and environmental dimension of the effects of climate change in South Africa and other emerging and developing economies, the contributions that business can make to climate change adaptation are an increasingly important area of investigation. The exploratory case study here is that of the South African retailer Woolworths, and its proactive engagement in water management. The retailer is one of only five South African companies listed on the Dow Jones Sustainability Index (DJSI). It was also one of only three companies in Africa to be named as one of the 16 sustainability champions of the developing world, in a report by the World Economic Forum and the Boston Consulting Group (Woolworths Holding, 2011b).

Business organisations can contribute to climate change adaptation by addressing risk areas within and beyond their operational boundaries. Yet, how companies respond is closely linked to why they act. The introductory chapter of this book (Chapter 1) highlighted the fact that a company's characteristics and strategic orientation influences how it responds to institutional drivers. Hence, this chapter aims to provide greater clarity on the interplay between institutional and organisational drivers that lead to voluntary efforts by business organisations to strengthen not only their own organisational resilience, but also that of the larger system in which they are embedded. Woolworths was established as a family business in 1931. Since the opening of its first store in Cape Town, it has become one of South Africa's most successful retailers, with over 400 stores throughout the country, and has started to expand into Africa and the Middle East (Luiz et al., 2011). The company focuses mainly on food, clothing and household goods, of which 90 per cent are sold under its private brand. Although Woolworths is quite small in comparison to its competitors, it holds a considerable market share, of about 30 per cent, in fresh produce.[1] Its target clientele are predominately medium- and high-income customers who want quality assurance, and who tend to be environmentally conscious (De Jager, 2009). Woolworths has built its brand differentiation on sustainability, high quality, consumer trust and innovation (Luiz et al., 2011). The focus on sustainability as a market differentiation has been very successful, as it has ensured the trust and loyalty of its suppliers and customers and, at the same time, has led to the company's international recognition (interview, J Smith, Sustainability Manager, Woolworths, 4 August 2011; Hamann et al., 2012).

Woolworths' climate change adaptation activities are led by its proactive engagement in water management and governance. The chapter explores what Woolworths has done to ensure sustainable water supplies, now and in the future, and considers the implications of this approach for managing water resources at the catchment and farm level. The focus on water is for two reasons. First, Woolworths has a large water footprint in its supply chain, particularly in the agricultural sector. Controlling one third of the national formal fresh produce market, it receives 95 per cent of its produce from South Africa. The company has therefore recognised that water will be one of its major risk areas under conditions of climate change (interview, J Smith, 4 August 2011). Second, given that the company has already implemented specific measures for addressing water as an area of risk, the focus on water is appropriate for illustrating the company's climate change adaptation strategy.

The author's initial interest in Woolworths was triggered by the observation that the retailer implements more far-reaching adaptation measures than, for example, its main competitor, Pick n Pay. Like Woolworths, Pick n Pay has identified climate change and poor quality and scarcity of water as risk factors (interview, A Nel, Senior Manager: Sustainable Development, Pick n Pay, 20 September 2011). However, Pick n Pay's measures for addressing these risks are still confined to its operations (Pick n Pay 2011; interview, A Nel, 20 September 2011). Preliminary findings showed that Woolworths is several steps ahead of its competitor. This is the case not only for issues around water management but also in regard to engaging with the topic of sustainability and its meaning in a fast-changing world where the environment, society and economy are interlinked in complex and non-linear ways.

The underlying research question that structured the research was therefore: What factors have led to Woolworths' extensive adaption strategy?

This chapter describes the methodology, which elicited rich knowledge on the company's adaptation strategy and the various drivers that may have contributed to it. The primary source of information was semi-structured interviews with representatives from Woolworths, the World Wide Fund for Nature (WWF), Pick n Pay and various experts in the field.[2] This was complemented by analysis of a diverse set of documents, as well as interview material from previous studies (Hamann et al., 2011).

A brief section on water governance in South Africa sets the context for understanding the role that business organisations such as Woolworths could potentially fulfil in an adaptive polycentric water governance system. Section three aims to understand how Woolworths contributes to climate change adaptation, by investigating the company's various water resource management measures and exploring their scope and scale, using the climate change adaptation typology described in the introduction of this book. Special attention is paid to the company's 'Farming for the Future' (FfF) programme, an 'environmental innovation' (Kemp and Pearson, 2007; Van den Bergh et al., 2011) that allows for system transformation (i.e. changes in relations and practices) and for building resilience within and beyond Woolworths by creating important links between different levels in the socio-ecological system.

Section four looks at *why* Woolworths is contributing to climate change adaptation, by considering the challenge of water governance in South Africa. Applying the concept of limited statehood (Börzel and Risse, 2010) to the empirical evidence, it seems that that the state's inability to deal with the significant problem of degrading water quality has prompted Woolworths to engage proactively in water resources management. However, it does not explain why Woolworths is expanding its sphere of influence beyond the firm level in a way that is unusual for its South African and international competitors. Examining Woolworths' organisational culture and its strategic orientation towards sustainability provides some answers to this question.

The chapter concludes by returning to the discussion of water management in the context of climate change, placing Woolworths' adaptation strategy in the larger water governance context.

Water management and governance in South Africa

South Africa is a water-stressed country, faced with environmental degradation, demographic pressures, large pockets of poverty, changing climate conditions and increased climate variability.[3] It thus requires a new approach to managing its water resources. While South Africa's progressive water legislation (RSA, 1998; DWA, 2004) strongly supports the move toward cooperative and integrated governance and stakeholder participation, the resources and skills needed to engage the multiplicity of stakeholders and facilitate communication and joint decision-making are still lacking. The current water governance system is therefore highly fragmented, and the government has limited capacity to coordinate the stakeholders who operate in the water sector and to protect South Africa's water resources. Consequently, water quality is continuously degrading and this has become a significant problem in most of South Africa's catchment areas. Furthermore, pollution as a result of inadequate municipal wastewater treatment plants, poor sanitation in informal settlements and unregulated discharge from industry and agriculture, as well as the infestation of alien invasive vegetation in the riparian zones, are serious threats to South Africa's water security (CSIR, 2010). Water governance in South Africa therefore features problems of limited statehood (Börzel and Risse, 2010).[4]

Woolworths' engagement in water management

This section examines the nature of Woolworths' engagement in water management. This is done by providing an overview of the three levels at which Woolworths has taken measures to contribute to sustainable water supply under current environmental conditions and future climatic changes. Measures are then classified according to the climate change adaptation typology proposed in the opening chapter. This typology makes it possible to link the different measures to three idealised adaptation types: coping, substantial adjustments and system transformation. Hence, it provides an indication of the scope and scale of Woolworths' adaptation strategy. Special attention is paid to the company's FfF programme, as this represents an 'environmental innovation' (Kemp and Pearson, 2007) that allows for system transformation and for building resilience within as well as beyond Woolworths. The latter is displayed by creating important links between different levels in the governance system. An emerging argument, which I will discuss in the later part of this section, is that each adaptation type seems to be facilitated by different learning processes. In the case of Woolworths, these learning capacities seem to be strengthened through Woolworths' polycentric governance approach – that is, its willingness to engage with other stakeholders in order to ensure adequate problem definition and problem solving on the issue of water security under conditions of climate change.

Understanding Woolworths' water management measures with the help of the adaptation typology

Although the following discussion will focus primarily on Woolworths' efforts in its supply chain through its FfF programme, it is important to note that the company is also active in water management at the operational level and at a level beyond its supply chain. Taking measures from all three levels into consideration demonstrates how different measures can be structured into the three adaptation types and further illustrates that these necessitate different sets of organisational capabilities.

Table 7.1 shows that the measures that Woolworths has taken within its operations and at the level beyond its supply chain can easily be defined as 'no-regret' measures, with the former leading to immediate cost savings and the latter contributing to corporate social responsibility. Business organisations now commonly use strategies of this kind, as they

Table 7.1 Woolworths' water management measures at three levels[5]

Level	Measures
Company operations (operational level)	• In-store water use monitoring via water meters • Water efficiency measures: e.g. rainwater harvesting and greywater recycling in new stores, underground spring utilisation at headquarters; real estate criteria include consideration of storm and wastewater management systems • Staff education: e.g. sensitising of buyers and technologists on water conservation , store campaigns
Supply chain (farm/sub-catchment level)	• Farming for the Future – agricultural model for sustainable water management (to be discussed in the text below)
Beyond the supply chain	• WWF's Water Balance Programme[6] – Sponsoring of alien clearing activities • Customer education through shopping bag campaigns and the company magazine *Taste* (e.g. water saving tips)

can be easily integrated into existing routines and practices and, if necessary, be reversed. Whether these measures are simply coping – defined here as measures that can be adjusted to existing routines and practices in order to reduce immediate impacts – or whether they trigger more substantial adjustments – shown by the development of new products or processes with the aim of reducing direct risks – depends on the motivation type (short-term vs long-term planning) and levels of investment into these measures. In the case of Woolworths, the interview data suggest that some of the measures taken at the operational level are part of a larger, emerging water management strategy that is characterised by long-term commitments, monitoring direct and indirect water use and introducing new products and processes. However, in their current form the activities are too limited in scope and scale to trigger significant changes in the company's practices.

Where Woolworths sets itself apart from other retailers is in its involvement in its supply chain (i.e. farm level). Through its FfF programme Woolworths has developed an innovative approach, through which it is able to engage proactively in water resources management at farm level. The 'FfF' programme may best be described as an 'environmental innovation' that allows for system transformation and for building resilience within and beyond Woolworths, by creating important links between different levels in the socio-ecological system. The term 'environmental innovation' has its origin in the business literature and refers to 'a product, production process, service or management or business method that is novel to the organisation (developing or adopting it) and which results, throughout its life cycle, in a reduction of environmental risk, pollution and other negative impacts of resource use (including energy use) compared to relevant alternatives' (Kemp and Pearson, 2007, 7). It is a response to an environmental problem that requires action. In this chapter it is used to refer to not only technological but also organisational and relational changes.

As outlined in the introductory chapter, and further elaborated in Chapter 6, system transformation is an adaptation type that refers to the restructuring of the interactions between different system components. To further illustrate that FfF is an action taken by Woolworths to restructure its relation with its supply chain in order to influence the wider system, a more detailed description of the programme is provided below.

FfF is a science-based agricultural model deployed at farm level that measures eight key sustainability components.[7] The majority of these have an impact, directly or indirectly, on water use and quality. The explicit reason for creating FfF was Woolworths' growing concern about how to ensure sustainable supply of its fresh produce. The company's technologists noticed that, despite technological and scientific advances in the agriculture sector, and despite increasing use of fertilisers and pesticides, yield and quality on their supplier farms had declined. At the same time, the company was growing quite substantially and thus requiring a larger supplier base (interview, K Pienaar, former Woolworths food technologist, 6 January 2012).

The aim of the FfF programme is to develop an alternative agricultural system, appropriate for the environmental conditions on South African farms, that uses soil health as its foundation. It is based on the understanding that, in the context of rising demand, existing environmental pressures and projected climatic changes, good quality produce is not achievable without good soil and water management – that is preserving the environmental integrity of the farming units. This approach emphasises water management in particular because sustainable land management practices depend on sustainable water management. Woolworths (Woolworths Holdings, 2012) has made a public commitment that by 2012 all fresh produce suppliers will be part of the FfF programme. In 2011 and 2012, the model is being expanded into horticulture, wine and dairy, with the intention that by 2015 50 per cent of the entire food business will be transformed (interview, J Smith, 4 August 2011).

That Woolworths engages in its supply chain through the programme in a novel way is best reflected by the dynamic, integrative and interactive nature of the approach. The programme was developed in collaboration with external experts to foster and verify, among others, sustainable land and water management practices on its supplier farms. It was launched in 2009 after three years of development based on scientific knowledge coming from Woolworths' in-house technologists, the environmental consultancy Enviroscientific and biodiversity experts. Farmers were actively involved in the overall approach. For example, Enviroscientific worked for an entire year with 40 supplier farmers (interview, K Pienaar, 6 January 2012). As part of the annual farm and pack house/processing facility audits that are intended to identify and prioritise risk areas at farm level, farmers are encouraged to explore possible causes jointly with the auditors, and to develop recommendations for improving or changing existing practices based on priorities made.[8] The model also fosters iterative learning processes in that it incorporates the knowledge gained from previous audits.[9] It furthermore takes the farm's context into consideration (its geographic and climatic area, the type of crop farmed, and so on) and can consequently detect diverse and newly emerging risks.

Since its inception, FfF has already had a measurable effect at farm level. According to the annual audits and external reviews (interview, L van Schoor, 7 March 2012), farmers have reduced the use of fertilisers and pesticides and are using more efficient, scientific methods of irrigation and management of wastewater. This has reduced not only the impact that farms have on the environment, but also the farmers' input costs (for water, fertilisers and electricity). These reductions in input costs, together with the improvement in the quality of their crops, have been an important selling point to the farmers. Rather than seeing FfF as a policing system or a checklist of imposed requirements (as it is often the case with international certification schemes such as GlobalGAP),[10] interviewees suggest that they value the expert advice provided by the auditor and the Woolworths technologists. To date, only one out of an estimated 140 suppliers has resisted implementing the FfF programme (interview, J Smith, 4 August 2011) and the farmers have shown great interest in the training workshops provided by Woolworths and the auditor (interview, L van Schoor, 7 March 2012).

To summarise, FfF is Woolworths' response to the growing problems of water insecurity. This response strategy is allowing Woolworths to interact in new ways with its supply chain and consequently in the catchment areas where its suppliers are located:

- It has given Woolworths a new way to engage in water resource management and governance. Instead of taking the approach traditionally used by business, of passing externalities on to the public, focusing exclusively on its own operations, or opting for international certification schemes (Porter and Kramer, 2011, Fuchs et al., 2009), Woolworths is willing to address an environmental problem that has its origin at farm level by providing information and technological expertise.
- It has changed relations along the supply chain. The annual audit and associated training allow for better knowledge transfer and skills development between Woolworths technologists, farmers and the auditing consultancy, Enviroscientific. The regular communication and information exchange between the three parties fosters learning processes and experimentation that in turn helps to improve the FfF programme.
- The audit includes compliance with environmental legal requirements and exerts pressure on the farmers to be aware of, and comply with, legislation that relates to their land and water management practice (e.g. registration of boreholes and end use, management of wastewater discharge, water abstraction licenses, removal of alien vegetation, etc.) In this way, the audit acts as an effective mechanism in the enforcement of existing state regulations.
- Finally, the programme has shifted the way both farmers and Woolworths think about productivity. For example, capacity building

at farm level has given farmers a more scientific view of the way productivity is directly linked to the environmental integrity of their land and water resources. Woolworths' food technologists, who previously focused mainly on food hygiene and safety, are now also emphasising sustainable land and water management practices – that is, they have a better understanding of the sources of the products that Woolworths is selling and how to manage corresponding resources.

Identifying learning capacities required for each adaptation type

Figure 7.1 shows that Woolworth has a fairly broad adaptation spectrum. Most measures at the operational level and beyond the supply level can be associated with coping strategies, with a few from the operational level qualifying as substantial adjustments. The investigation into Woolworths' efforts in its supply chain showed that its FfF programme can be related to system transformation. The information summarised in Figure 7.1 also implies that each adaptation type requires a different set of organisational capabilities. Findings from the investigation into the FfF programme further suggest that influencing changes in the larger system necessitates greater interaction with other actors operative in the system. Yet, moving up the continuum of adaptation types (associated with broader scope and scale), demands not only technical skills and financial resources, but also the questioning of existing practices as well

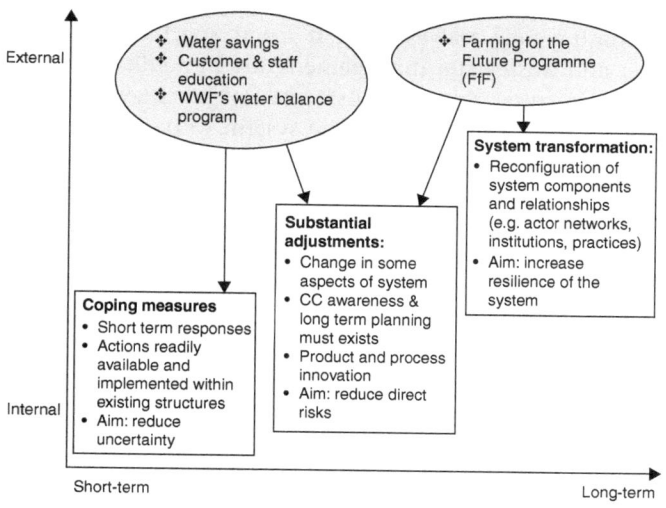

Figure 7.1 Woolworths' adaptation spectrum in the water sector

as the underlying assumptions and values on which these practices are based. Hence, it requires specific learning capacities.

To illustrate this further it will be helpful to link the different adaptation types, and the learning processes that facilitate them, to the concepts of double- (Argyris and Schön, 1978) and triple- (Hargrove, 2002) loop learning. Single-loop learning implies improving strategies and routines without questioning their underlying assumptions, whereas double-loop learning involves examining assumptions (e.g. cause–effect relationships), and triple loop learning revolves around reconsidering underlying values, beliefs and worldviews. What needs to be highlighted in this discussion is that each learning type will lead to different problem-definitions and -solving when confronted with complex issues such as water security under conditions of climate change.

Coping is mostly a result of single-loop learning, as it only involves making incremental changes to achieve predefined management goals, without questioning underlying assumptions. Here the goal is, for example, to reduce Woolworths' existing water consumption by 30 per cent, which might be achieved via water saving devices. Substantial adjustments, on the other hand, spring from reflecting on how the problem is framed and whether the goal is appropriate, and this reflection leads to innovative approaches and new roles, processes and institutions. This involves a process of reflecting on what implications water, as a risk factor, has for the retailer. Utilising alternative water sources (e.g. greywater recycling and tapping into an underground spring at the headquarters), and purchasing new real estate if it meets specific criteria for effective water and wastewater management, are more likely to result from double-loop learning. System transformation is achieved by examining assumptions about the nature of the system to be managed. The outcomes of such higher learning processes are often that new boundaries are set and new actor-networks established, leading to a redefinition of underlying values and norms (see for example Pahl-Wostl et al., 2007; Pahl-Wostl, 2009; Huntjens et al., 2011). It seems that FfF, with its specific focus at the farm level, was only possible because the company questioned the very foundation of water security and their need to intervene at system level (interview, J Smith, 4 August 2011). In turn, FfF led to new relations and boundary setting as well as to genuinely questioning 'why are we doing what we are doing and how can it be changed' (interview, K. Pienaar, 6 May 2011).

Findings from the water governance (e.g. Pahl-Wostl, 2007; Pahl-Wostl, 2009) and climate change literature (e.g. Pelling et al., 2008) seems to indicate that higher processes of learning (such as double- and triple-loop learning) rely on informal learning processes and the inclusion of different knowledge sources. Woolworths' collaborative partnership with the conservation organisation WWF SA exemplifies how the company strategically engages other stakeholders on issues of climate change and water management.[11] According to the Woolworths employees who were interviewed, the trusting partnership that has evolved over the years has been very beneficial for the company, helping managers to better understand the emerging risk landscape in which the company operates. WWF's expertise on climate change, biodiversity and water management has been of particular value for Woolworths, as it has helped to reveal hidden risks in its supply chain. Furthermore, WWF's endorsement of initiatives such as FfF improves Woolworths' reputation. In turn, WWF collaborates and shares knowledge with Woolworths because it knows that big retailers are in a powerful position to influence suppliers and as well as customers. It seems that Woolworths' polycentric approach for addressing the issue of water security has resulted not only in a better understanding of the risk landscape and a more encompassing problem definition but also in greater problem-solving capacities. In particular Woolworths' collaborative efforts (i.e. investment in relational capacities) has built resilience within Woolworths, by improving managers' ability to detect risks and develop in-house expertise to deal with them, and beyond Woolworths, by building the capacity of supplier farmers.

Institutional and organisational drivers that shape Woolworths' adaptation spectrum

The question remains why Woolworths has such an extensive adaptation spectrum – in other words, why the company chose to engage in water management beyond its internal operations. and what factors enabled it to do so. In this context, I consider several institutional drivers and highlight specific organisational drivers as well as company characteristics that may have contributed to Woolworths' response to the climate change.[12] The findings suggest that the prevailing governance gap in the water sector (institutional driver) and the retailer's branding around the notion of sustainability (organisational driver) explain why Woolworths engages proactively. Its organisational culture and long-term relationships with its suppliers, on the other hand, enable proactive engagement and explain the extent of this engagement.

Institutional drivers: Pressures and incentives

The introductory chapter suggested that state regulations can be compelling incentives for companies to contribute to environmental governance. However, in South Africa, an emerging economy confronted with many social and environmental problems, state capacity to enforce regulations is limited. The issue of degrading water quality in many South African catchments is an indicator for the state's limited capacity to deal effectively with the pressures that threaten South Africa's water resources. Currently the state cannot regulate and coordinate actions, either through hierarchical governance or through a non-hierarchical mode involving both state and non-state actors. This then raises the question of what other factors have compelled Woolworths, a profit-maximising private actor providing private goods, to invest in the management of water outside its operational boundaries and without necessarily reaping immediate short-term benefits.

Woolworths is well aware that the government will not be able to manage the looming water crisis on its own, nor does it expect that the state will take on leadership to ensure engagement of the various actors involved in the water sector in the near future. A passive 'wait and see' approach is therefore not an option for securing its production base in the long-term. Justin Smith explains Woolworths' position thus: 'Business leadership for sustainability requires not waiting for the government to make changes, but rather going forward – government will catch up' (email interview, J Smith, 23 December 2011). Lack of confidence in the state can to some extent explain Woolworths' willingness to step outside its traditional boundaries and make innovative contributions to water governance. In countries with consolidated statehood, a retailer with a profile similar to Woolworths' (e.g. Marks and Spencer in the UK) would most likely be less incentivised to contribute to water governance that goes beyond its operations, as it would expect the state to deliver the public good.

As a very visible brand serving a high income, environmentally and socially conscious consumer base, normative structures such as environmental and human rights norms also affect Woolworths' adaptation behaviour. These norms are important not only to public actors or NGOs but also to customers (see Hall, 2000). This is especially true for those high- and medium-income customers who care about the environmental and social implications of the products they buy and the brands they support. As agued by the company's chairperson, Woolworths has always found it important to engage proactively with these matters in order to strengthen its reputation and legitimacy (interview, S Susman, 26 January 2012). With water scarcity and degrading quality featuring prominently in the South African media, and increasingly in political debates, a Woolworths' employee pointed out that: 'we don't want to find ourselves in a situation where our customers turn to Woolworths asking what Woolworths is doing to address the water issue and we don't have a satisfactory answer for them' (interview, T McLaughlin, 13 April 2011).

The discussion suggests that the prevailing governance challenges around water and normative pressures have enhanced Woolworths' willingness to take a more proactive approach toward water management. Yet, on their own, these institutional drivers do not suffice to persuade a company to advance adaptation strategies that require major changes in its core business practices. To better understand Woolworths' leadership role, and the way in which its FfF programme was able to materialise, the author turns to several organisational drivers that are of significance to the company's strategic orientation, namely its branding strategy, organisational culture, and long-term relationship with its suppliers. Together these company characteristics represent key elements of Woolworth's polycentric governance approach and overall enhance its learning capacities.

Organisational drivers: Branding strategy, organisational culture and supplier relationships

Woolworths' approach to climate change adaptation cannot be explained without considering its long-standing branding around the notion of sustainability. The company, which has built its brand on the concept of high quality at an affordable price, seeks to differentiate itself from other retailers by emphasising that it is a brand that cares and that it 'must always match the future expectations of [its] customers' (interview, S Susman, January 2012).[13] According to Simon Susman (interview, January 2012) Woolworths' orientation towards sustainability started in the 1950s through its strong focus on good quality and 'value for money'. Back then, the company realised it needed to develop strong relationships with its suppliers to ensure the high quality of its products. In the 1990s Woolworths developed its 'good food journey', which signified going back to the source of the product. In this phase, the importance of environmental sustainability became evident.

In 2007 Woolworths institutionalised this commitment through the launch of its Good Business Journey (GBJ), a five-year sustainability plan (Woolworths Holdings, 2007). The aim of the GBJ is to keep improving the company's sustainability performance in all of its business units as well as along the supply chain (Woolworths Holdings, 2008).[14] By setting specific targets, and using a sophisticated tracking system (using scientific indicators to measure performance) to monitor and evaluate progress towards these targets, Woolworths has created an evidence-based management approach.[15] This helps to provide greater legitimacy for far-reaching adaption measures such as the FfF programme, provided that they can be shown to enhance Woolworths' corporate strategy. Hence, it helps to reduce potential internal resistance to substantial organisational changes or investments that may be required.

The investigations also revealed that Woolworths' organisational culture, in particular the complementary leadership roles between top management (in particular the former CEO Simon Susman) and mid management have been a device factor in facilitating the necessary organisational change (interview, J Smith, 4 August 2011; interview, K Pienaar, 6 January 2012). This allowed for sustainability champions to grow within the organisation and to influence the company's policies. This is exemplified by the FfF programme, which required considerable risk, investment and change in practices. The programme was a bottom-up initiative that started as a pilot scheme. Under the leadership of a food technologist, several of those mid-management champions transformed the idea of a new agriculture system based on soil fertility into a business model. Once these internal activists could show how the programme links to and enhances Woolworth's strategic orientation it was endorsed by top management, and it is now implemented across the organisation (interview, S Susman, January 2012). Additionally, these champions give Woolworths access to the resources and capacities of other networks and sustainability experts to which they are connected.

Another defining feature of Woolworths' corporate strategy that has enabled Woolworths to step outside its traditional boundaries and engage in water governance beyond the company level has been its direct and long-term relationship with its supplier farms. The involvement of independent scientists who measure continual improvement and prioritise risks at farm level adds value and improves Woolworths' relationship with suppliers. Conceptually, this approach is supported by Dyer and Singh (1998), who investigate the way relational investments contribute to and redefine competitive advantage and observe that relation-specific investments – such as long-term relationships with suppliers – lead to cost reductions in the value chain and allow for collaboration and cooperation in a fast-changing environment.

The convergence of these organisational drivers and specific institutional drivers is perhaps best illustrated by comparing Woolworths' approach to that of both Pick n Pay and the British retailer Marks and Spencer. Being a South African retailer, Pick n Pay is confronted with the same institutional drivers as Woolworths (e.g. it operates in an area of limited statehood). Yet, Pick n Pay lacks some of the specific organisational drivers and capacities that made it possible for Woolworths to develop FfF, and to integrate it successfully into its supply chain. Marks and Spencer, meanwhile, has a similar approach towards sustainability, and pursues a branding strategy comparable to that of Woolworths.[16] Marks and Spencer is not challenged by limited statehood in its home country, and this might explain why it has not developed an environmental innovation similar to that of FfF. Marks and Spencer's supply chain, however, is operative to some extent in countries where the state is too weak to enact or enforce environmental legislation (e.g. Marks and Spencer's has fresh produce suppliers in Kenya). It might be for this reason that the UK retailer has recently approached Woolworths for advice on to the development of a programme similar to that of FfF.

Conclusion

The research described in this chapter discusses issues of how and why business organisations contribute to climate change adaptation in ways that strengthen not only their organisational resilience but also that of the larger system in which they are embedded. This was done through an exploratory case study of the South African retailer Woolworths and its proactive engagement in water management.

The findings show that Woolworths' engagement beyond its internal operations derives from the convergence of specific institutional and organisational drivers. Cognisant of the state's challenges in coordinating water management effectively and the projected climatic change impacts for South Africa, Woolworths is willing to take a proactive approach, focusing on future water supply risks. However, the existing institutional pressures coincided with a number of specific organisational drivers of which some motivated and other enabled the company's proactive engagement with water resources management. Woolworths' branding around the notion of sustainability, its organisational culture and its direct and long-term relationship with suppliers are specific characteristics of its corporate strategy that have played a key role in the company's climate change adaptation approach and in its proactive engagement in water management.

The analysis further suggests that the interplay between organisational and institutional drivers, leading to its proactive engagement at farm level, is based on Woolworths' specific learning capacities. Woolworths was able to cultivate these learning capacities mainly through its investments into relational capacities (not only within its own operations but also with suppliers, customers and experts). The regular information exchange with external experts and suppliers through its polycentric governance approach and the resulting mutual learning processes have made it possible for Woolworths to question existing practices and their underlying assumptions. The higher learning processes resulted in organisational changes: the implementation of FfF can be explained by existing organisational leadership and the long-term relationship with suppliers. Whereas the former minimised internal resistance the latter reduced resistance in the supply chain. Overall the findings suggest that by taking a polycentric governance approach and linking its climate change adaptation approach in the water sector closely to the socio-ecological system in which it is embedded, Woolworths has strengthened its internal capacity to deal with hidden and emerging risks.

In essence, Woolworths actively contributes to the polycentric governance that seems to be needed to achieve sustainable water management under conditions of existing environmental changes and future climate change (Pahl-Wostl et al., 2011a; Newig and Fritsch, 2009). In the absence of an effective central authority (i.e. the state) the strengthening of polycentric governance arrangements has been highlighted as a promising approach for dealing with collective action problems[17] such as water management under conditions of climate change (Pahl-Wostl et al., 2011b; Huitema et al., 2009). The term 'polycentricity' (Ostrom et al., 1961) refers to 'many centres of decision making that are formally independent of each other [...] to the extent that they take each other into account in competitive relationships, enter into various contractual and cooperative undertakings or have recourse to central mechanisms to resolve conflicts' (Ostrom, 2010: 552). The advantage of effective polycentric systems is that they are able to balance bottom-up and top-down processes and establish cooperative links between the different centres of decision-making (see e.g. Pahl-Wostl, 2009; Pahl-Wostl et al., 2012) and thus 'tend to enhance innovation learning, adaptation, trustworthiness, levels of cooperation ... and the achievement of more effective, equitable, and sustainable outcomes at multiple scales' (Ostrom, 2010: 552). The empirical findings, as discussed earlier, show that Woolworths promotes this type of polycentric governance on an internal as well as external scale. This significantly enhances the retailer's adaptive capacity[18] and learning capacities. This in turn enables Woolworths to take its adaption efforts beyond incremental adjustments and to contribute in compensating the current governance gap.

Woolworths' engagement in its supply chain shows that Woolworths fulfils some important governance functions. Woolworths has been able to exert its influence within, as well as across, various levels and uses this influence within its operations and along the supply chain to promote sustainable water management practices. Yet one must not overestimate the retailer's willingness to fulfil such a role once it requires efforts not directly linked to its core business (in this case its fresh produce supply). To enhance the company's willingness to engage even further with other actors and decision-making processes that are important for the management of South Africa's water resources would only be possible by demonstrating a link to its core business. This is manifested in management's current belief that the company should not play a greater role at catchment scale (e.g. via engagement with and support of catchment agencies) but rather that it should restrict its influence to the farm level.

Appendix 1   List of interviews

Name	Affiliation (position)	Dates of interviews
Tom McLaughlin, Johan Ferreira, Kobus Pienaar	Woolworths, respectively, Manager: Good food journey; Manager: Food business unit; Food Technologist	23 July 2008
Tom McLaughlin	Woolworths, Manager: Good food journey	1 August 2008 13 April 2011 24 February 2012
Mao Amis	WWF SA, Manager: Integrated Catchment Management	3 May 2011
Rodney February and Helen Gordon	WWF SA, Water Balance Program Managers	3 May 2011
Kobus Pienaar	Woolworths, (formerly) Food Technologist	6 May 2011 6 January 2012
Tatjana von Borman	WWF SA, Market Transformation Manager	3 August 2011
Justin Smith	Woolworths, Sustainability Manager	4 August 2011, 11 November 2011 23 December 2011 email interview
Andre Nel	Pick n Pay, Senior Manager: Sustainable Development	20 September 2011
Simon Susman	Chair of the Board of Directors Woolworths, CEO from to 2000–2010	26 January 2012
Lourens van Schoor,	Enviroscientific	7 March 2012

Notes

1. This 30 per cent does not take into account South Africa's large informal fresh produce market.
2. See Appendix 1 for complete list of interviews.
3. The IPCC's global climate model predicts that the eastern and central regions of South Africa will become wetter and the western and southern regions drier and

hotter (Stuart-Hill and Schulze, 2010). The predicted changes in rainfall season-ality and variability, and the rise in temperature, will alter the hydrological sys-tems that determine the availability of water in South Africa (Schulze, 2005).

4. The concept area of limited statehood refers to a situation (geographical or thematic) where the state is not able or willing to enforce rules or control the means of violence (Börzel and Risse, 2010).

5. Please note that this table is for illustrative purposes only and does not pro-vide a complete list of all of the company's water-related measures/activities.

6. Woolworths has been participating in WWF's Water Balance Programme, which assists companies to become water stewards by voluntarily monitor-ing and reducing the use of water in their operations and by increasing the supply in the overall system through the sponsoring of alien clearing (interview, R February and H Gordon, WWF SA, Water Balance Programme Managers, 3 May 2011).

7. These eight main components are soil management, irrigation water man-agement, environmental legal requirements, biodiversity management, waste and wastewater management, cooling and energy use and carbon foot-print, pest and plant management (all have measurable sub-categories) as well as a self audit by the farmer (interview, L. van Schoor, 7 March 2012).

8. The risk assessments on the farms have shown that most farmers tend to over irrigate and do not manage wastewater properly. It is not that they intention-ally pollute the river, draw more water than they need or violate existing legislation. Rather, they are often not aware of the various, sometimes not obvious, ways that their practices damage the water resources (rivers and groundwater) on which they depend (interview, K. Pienaar, 6 May 2011).

9. Woolworths pays for the audit but the farmer has to cover the cost of the measures to be implemented.

10. GlobalGAP stands for the Global Partnership for Good Agriculture Practice. It is an international certification scheme. So far, requirements of schemes such as GlobalGap have been without an adequate scientific foundation, and they do not necessarily assist in the move towards sustainable land and water management practices (see, e.g. Fuchs et al., 2009).

11. Woolworths has engaged in both contractual and informal alliances with experts such as the CSIR (Council for Scientific and Industrial Research), the Graduate School of Business and ENGO at the University of Cape Town, the environmental consultancy Enviroscientific, the consultancy Pegasus and the GIZ (Agency for International Cooperation, Gesellschaft fuer Internationale Zusammenarbeit, formerly GTZ, German Technical Cooperation) on issues of climate change and water management. I focus here on the partnership with WWF SA as it best illustrates the emphasis on mutual learning.

12. The author understands internal factors that motivate a company to do certain things as organisational drivers, whereas company characteristics are closely related to the capacities of the company that allow it to implement its response. Yet the distinction between the two is quite difficult, as a driver can also be a capacity. For example, Woolworths' relationship with suppliers can be seen as a driver, because it provided a rationale to ensure long-term productivity on these farms. Yet it is also a capacity. Without this relationship, the company could not have implemented the FfF programme feasibly. For the remainder of the chapter I therefore will broadly refer to them as company drivers.

13. Börzel et al (2010) state that companies like Woolworths, which cater for high-end markets and specialise in brand products (i.e. private brands) pursue sustainability efforts via strict self-regulation to increase the market value of their products.
14. Of the targets listed under the priority area 'Environment', several have implications for Woolworths' commitments in the realm of water management: i. Reduce Woolworths' water consumption by 30 per cent, ii. Enforce a strict code of conduct regarding ... water management in the supply chain, and iii. ensure that conventional produce farmers migrate to organic production or environmentally sensitive farming methods (Woolworths Holdings, 2011a; 2011c). The water management measures discussed in Section 3 form an integral part of the GBJ, as they are responses to these three targets.
15. The system entails over 200 indicators, each with a one-year and a five-year target, providing executive management with an accessible overview of how the company is performing relative to targets, in particular in thematic areas or specific business units (Hamann et al., 2012).
16. See Marks and Spencer's Plan A which is comparable with Woolworth's GBJ (Marks and Spencer, 2012).
17. Collective action problems have been associated with common pool and open access resources where individual actions will only lead to suboptimal outcomes as such problems require the collaboration of all actors involved.
18. Adaptive capacity is here defined as the ability of the system to alter processes and if required convert structural elements (Pahl-Wostl, 2009). More specifically it refers to a set of tangible (e.g. economic capital, physical resources and human capital) and less tangible (e.g. social capital including relational capacities) resources as well as the ability to mobilise these resources (Nelson et al., 2007). Hence, it sets the parameter for learning and innovation.

References

Argyris, C. and Schön, D. (1978) *Organizational Learning: A Theory of Action Perspective* (Reading, MA: Addison Wesley).

Börzel, T.A. and Risse, T. (2010) 'Governance without a State: Can it work?', *Regulation and Governance* 4, 2, 113–134.

CSIR (2010) A CSIR perspective on water in South Africa – 2010. CSIR Report No. CSIR/NRE/PW/IR/2011/0012/A ISBN: 978-0-7988-5595-2.

De Jager, J.W. (2009) *Profiling SA's Green Consumer*. Unpublished Masters dissertation, University of Cape Town Department of Environmental and Geographical Science.

DWA (Department of Water Affairs and Forestry RSA) (2004) *National Water Resource Strategy. Our Blue Print for Survival* (1st ed.) (Pretoria: South Africa).

Dyer, J.H. and Singh, H. (1998) 'The Relational View: Cooperative Strategy and Sources of Interorganizational Competitive Advantage', *Academy of Management Review* 23, 4, 660–679.

Fuchs, D., Kalfagianni, A. and Arentsen, M. (2009) 'Retail Power, Private Standards, and Sustainability in the Global Food System', in J. Clapp and D. Fuchs (eds), *Corporate Power in Global Agrifood Business* (MIT Press), pp. 29–60.

Hall, J. (2000) 'Environmental Supply Chain Dynamics', *Journal of Cleaner Production* 8, 445–471.

Hamann, R.,. Giamporcaro, S., Johnston, D. and Yachkaschi, S. (2011) 'The Role of Business and Cross Sector Collaboration in Addressing the "Wicked Problem" of Food Security', *Development Southern Africa* 28, 4, 579–594.

Hamann, R., Methner, N. and Nilsson, W. (2012) *The Evolution of a Sustainability Leader: The Development of Strategic and Boundary Spanning Organizational Innovation Capabilities in Woolworths*, EURAM '12 Conference, 6– 8 June 2012, Rotterdam.

Huntjens, P., Pahl-Wostl, C., Rihoux, B., Schlüter, M., Flachner, Z., Neto, S., Koskova, R., Dickens, R., Nabide, I. and Kiti, N.I. (2011) 'Adaptive Water Management and Policy Learning in a Changing Climate: A Formal Comparative Analysis of Eight Water Management Regimes in Europe, Africa and Asia', *Environmental Policy and Governance* 21, 145–163.

Hargrove, R. (2002) *Masterful Coaching*. Revised edition (Jossey Bass).

Huitema, D., Mostert, E., Egas, W., Moellenkamp, S., Pahl-Wostl, C. and Yalcin, R. (2009) 'Adaptive Water Governance: Assessing the Institutional Prescriptions of Adaptive (co-) Management from a Governance Perspective and Defining a Research Agenda', *Ecology and Society* 14, 1, 26. [online] URL: http://www.ecologyandsociety.org/vol14/iss1/art26/ [accessed 17 November 2011].

Kemp, R. and Pearson, P. (eds) (2007) Final Report MEI Project about Measuring Eco-Innovation. (MERIT, University of Maastricht: Maastricht). http://www.merit.unu.edu/MEI [Accessed 8 September 2011].

Luiz, J., Bowen, A. and Beswick, C. (2011) *Woolworths South Africa: making sustainability sustainable*. Emerald Emerging Markets Case Studies Collection Vol. 1 No. 1: 1–21. DOI: 10.1108/20450621111113534.

Marks and Spencer (2012) *Your M and S How We Do Business Report 2012*. Available at: http://plana.marksandspencer.com/media/pdf/ms_hdwb_2012.pdf [Accessed 15 February 2012].

Nelson, D.R., Adger, W.N. and Brown, K. (2007. 'Adaptation to Environmental Change: Contributions of a Resilience Framework', *Annual Review of Environment and Resources* 32, 395–419.

Newig, J. and Fritsch, O. (2009) 'Environmental Governance: Participatory, Multi-level, and Effective?', *Environmental Policy and Governance* 19, 3, 197–214.

Ostrom, E. (2010) 'Polycentric Systems for Coping with Collective action and Global Environmental Change', *Global Environmental Change* 20, 550–557.

Ostrom, V., Tiebout, C.M. and Warren, R. (1961) 'The Organization of Government in Metropolitan Areas', *American Political Science Review* 55 (Dec.), 831–842.

Pahl-Wostl, C. (2007) 'Transitions Towards Adaptive Management of Water Facing Climate and Global Change', *Water Resources Management* 21, 49–62.

Pahl-Wostl, C. (2009) 'A Conceptual Framework for Analyzing Adaptive Capacity and Multi-level Learning Processes in Resource Governance Regimes', *Global Environmental Change* 18, 354–365.

Pahl-Wostl, C., Sendzimir, J., Jeffrey, P., Aerts, J., Berkamp, G. and Cross, K. (2007) 'Managing Change toward Adaptive Water Management through Social Learning', *Ecology and Society* 12, 2, 30. www.ecologyandsociety.org/vol12/iss2/art30/ [accessed 15 May 2009].

Pahl-Wostl, C., Nilsson, C., Gupta, J. and Tockner, K. (2011a) 'Societal Learning Needed to Face the Water Challenge', *AMBIO* 40, 549–553.

Pahl-Wostl, C., Lebel, L., Knieper, C. and Haeyer, T.D. (eds) (2011b) Synthesis Report. Context-Sensitive Comparative Analysis of Associations between Water Governance Properties and Performance in Water Management. Twin2Go Deliverable No. 2.3.

Pahl-Wostl, C., Lebel, L., Knieper, C. and Nikitina, E. (2012) From applying panaceas to mastering complexity: Toward adaptive water governance in river basins Environmental Science & Policy 23 (2012), 24–34.

Pelling, M., High, C., Dearing, J. and Smith, D. (2008) 'Shadow spaces for social learning: a relational understanding of adaptive capacity to climate change within organisations', *Environment and Planning* 40, 4, 867–884.

Pick n Pay (2011) Company website. www.picknpay.co.za [accessed 23 November 2011].

Porter, M.E. and Kramer, M.R. (2011) 'The Big Idea. Creating Shared value: How to Reinvent Capitalism and Unleash a Wave of Innovation and Growth', *Harvard Business Review* Jan–Feb. 2011.

RSA (Republic of South Africa) (1998) National Water Act. Act No. 36 of 1998, Government Gazette. South Africa.

Schulze, R.E. (2005) 'Setting the Scene: The Current Hydroclimatic "Landscape" in Southern Africa', in R.E. Schulze, (ed.) *Climate Change and Water Resources in Southern Africa: Studies on Scenarios, Impacts, Vulnerabilities and Adaptation* (Pretoria: Water Research Commission. WRC Report 1430/1/05: Chapter 6, 83–94).

Stuart-Hill, S. and Schulze, R.E. (2010) 'Does South Africa's Water Law and Policy Allow for Climate Change Adaptation?', *Climate and Development* 2, 2, 128–144.

Van den Bergh, J., Truffer, B., and Kallis, G. (2011) 'Environmental Innovation and Societal Transitions: Introduction and Overview', *Environmental Innovation and Societal Transitions* 1, 1–23.

Woolworths Holdings (2007) Woolworths announces the 'Good business journey', Press Release.19 April 2007. www.woolworthsholdings.co.za/media/news/news_display.asp?Id2=83 [accessed 13 January 2012].

Woolworths Holdings (2008) Annual report 2008. Available via: http://www.woolworthsholdings.co.za [Accessed 20 November 2011].

Woolworths Holdings (2011a) Annual report 2011. Available via: http://www.woolworthsholdings.co.za [accessed 17 January 2012].

Woolworths Holdings (2011b) Woolworths is recognized as a global leader in sustainability. Press release 16 September 2011. http://www.woolworthsholdings.co.za/media/news/news_display.asp?Id2=486 [accessed 15 November 2011].

Woolworths Holdings (2011c) Good Business Journey Report 2011. Available via: http://www.woolworthsholdings.co.za [accessed 17 January 2012].

Woolworths Holdings (2012) Company website: www.woolworths.co.za [accessed 13 January 2012].

8
Insurance, Climate-Risk and the Barriers to Change

Tom Herbstein, Jan Froestad, Deon Nel and Clifford Shearing[1]

Introduction

The cost of extreme weather events, resulting from climate change, has increased by 37 per cent per decade since the 1980s (Mills, 2009). This has raised serious questions about the financial sustainability of the insurance industry and, indeed, about its future role as the world's risk manager, prompting the industry to begin to reflect on how it might better manage the impacts of climate risk in the future.

The direct link between climate change and insured losses has led several commentators, starting with Leggett (1993), to propose that the insurance industry should be motivated to shift its practices to support the mitigation of climate change. Leggett lobbied insurers to strategically disinvest their significant financial assets away from the most polluting industries, thereby influencing the release of greenhouse gases (GHG) over the long-term.

But there has been little evidence of insurers responding in the way Leggett proposed. Instead, they have focused mainly on improving their ability to assess climate risk and avoid its impacts, rather than actively mitigating the drivers behind climate change. Part of the reason for this, it has been suggested, is that the threat of short-term losses to their investment portfolios may outweigh any benefits that could stem from mitigating climate change over the long term (Paterson, 2001; Phelan et al., 2011).

In this chapter, we consider a recent (2010–11) study, undertaken by a well-established South African short-term insurer that is beginning to see the impact of climate risk affecting its own business activities. The study examined the Eden District Municipality, located to the east of Cape Town on South Africa's southern Cape coast.

Between 2003 and 2008 the area was affected by a series of droughts, floods, storm surges and wild fires, costing the region an estimated R2.5 billion (US$300 million) in total economic losses (Holloway et al. 2010) and accounting for almost 80 per cent of all South Africa's catastrophic insurance losses over the same period.

The study was initiated to provide The Santam Group (Santam) with a better understanding of what shaped the risk landscape in the Eden District, with the hope that this would enable Santam to begin to better understand the exposure of insured assets and thus help reduce its losses.

The study highlighted how part of the challenge for insurers is their continued reliance on risk assessment, especially actuarial analysis, as the primary tool for managing exposure to climate risk. Although this has been an effective tool for helping to maintain sustainable levels of exposure in the past, actuarial analysis is based on the assumption that past losses are indicative of future trends – a view climate change is beginning to challenge.

But part of the problem for insurance companies is that they have not always fully recognised the influence that other physical features within the local environment (such as land-surface hardening and deforestation), besides a changing climate, have in contributing to the size and impact of a climate risk (such as a flood). The study highlighted the influence, both directly and indirectly, that insurers potentially have over institutions that are able to contribute to shaping these 'proximate drivers' of climate risk. This offers insurers the opportunity to manage more proactively the exposure of insured assets.

However, in order to understand why insurers have focused on improving their capacity to assess, rather than address climate risk, it is necessary to identify why companies often respond to the need for change by focusing on improving their existing solutions – or ways of doing – first, before developing new ones (March, 1991; Moser and Ekstrom, 2010). Part of the reason for this is that innovation often requires high levels of risk-taking and uncertainty. This is certainly the case with regard to climate risk, where insurers need to reinvent the way they manage risk, a challenge often exacerbated by insurers' pre-existing business mentalities, the short-term nature of the industry, competition pressures and the constraints posed by the broader regulatory environments within which they operate.

A focus on risk assessment

Insurance is the largest industry in the world – bigger than the defence, oil, electricity generation or pharmaceutical industries (Phelan et al.,

2011). It represents seven per cent of global GDP, with commercial insurers alone turning over $4.6 trillion in annual premiums in 2010 (The Geneva Association, 2011). With financial investments valued at $24.6 trillion, insurance is also the third largest institutional investor, following pension funds ($29.9 trillion) and mutual funds ($24.7 trillion) (Maslakovic, 2011).

To ensure its own resilience and protect that of the broader pool of assets it underwrites, insurance limits its exposure to risks that are both *sudden and unexpected*, regarding more predictable losses as those that should be addressed by a client's own prudent risk-management strategies. The most important tool for managing insurers' risk exposure in the past has been its use of actuarial analysis – the use of past claims data to predict the probability of future losses. Actuarial analysis has proven to be a highly reliable and effective tool in helping insurance manage its exposure to risk in the past.

Yet insurance today is faced by a far more complex and interconnected world, which threatens to undermine its ability to calculate effectively the probability of some risks. Of these, climate change is arguably the most worrying, having emerged as one of the primary threats facing the world this century (Millenium Ecosystem Assessment, 2005; IPCC, 2007: 5). With temperatures estimated to rise by 0.2 °C per decade over the coming century, the impact this will have on the Earth's natural systems will be profound and likely to compromise the natural buffering capacity of the Earth's ecosystems severely (Millenium Ecosystem Assessment, 2005). The direct and indirect impacts of climate change will cost the global economy anywhere between 5 and 20 percent of GDP, depending on the rate at which future GHG emissions continue to rise (Stern, 2007).

This deterioration in the environment has been further compounded by the weakness of governance systems in establishing the necessary rules to address climate change (Gunningham et al., 1998; Cole and Grossman, 1999; Durant et al., 2004; Holley et al., 2011; Farrier and Whelan, 2004; Börzel and Risse, 2010; Freeman and Farber, 2005; Gunningham and Sinclair, 2002; Porter and Kramer, 2002). These challenges are compounded in areas where state-centred regulation and enforcement is already weak (Börzel and Risse, 2010).

The regulatory capacity of insurance

Since the 1980s, insured losses linked to 'natural' disasters, have been rising sharply (Mills, 2009). Figure 8.1 emphasises the rate at which

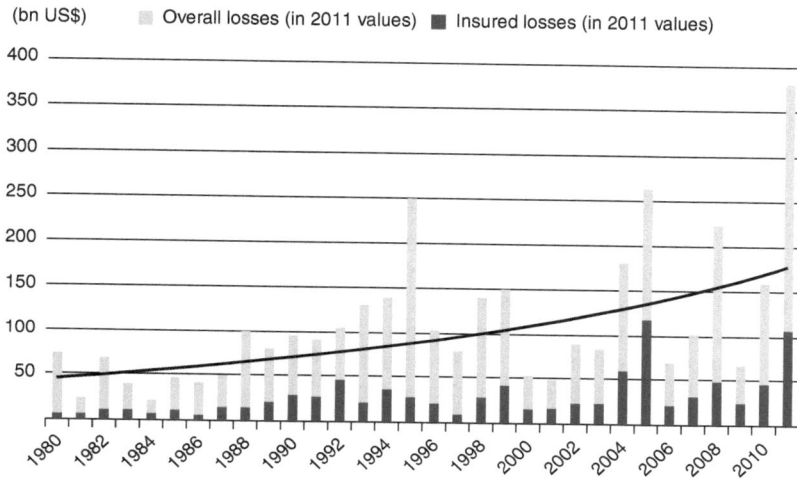

Figure 8.1 Trends in insured losses due to natural catastrophes and man-made disasters 1980 to 2011 (Munich Re, 2012)

losses for the insurance industry have risen over the past 30 years. Two features that stand out are the steady growth of the 10-year moving average and the increase in individual annual losses, both in frequency and intensity, since 1990. Five of the costliest years on record have occurred over the past decade. Although 2011 was the costliest year yet in terms of total economic losses, Japan not being 'fully insured' against tsunami loss (Business Week, 2011), this leaves 2005, and the $123 billion losses attributed to hurricanes Katrina, Wilma and Rita, as the costliest yet for the global insurance industry. However, future predictions indicate that extreme year payouts for insurance could well exceed $1,000 billion (Mills, 2005).

In this light, insurance has been described as the industry with the most to lose as a result of climate change (Leggett, 1993; Gelbspan, 1998), a feature that was enough to motivate the industry to begin to address its exposure more proactively. Early attempts to do this date back to 1992, when Jeremy Leggett, then head of Greenpeace, sought to encourage insurers to engage more in global climate politics (Paterson, 2001). Leggett strongly promoted efforts to mitigate global GHG emissions by lobbying insurers to strategically disinvest their financial assets away from the most polluting industries (Leggett, 1993). He hoped this would help to address the primary drivers behind climate change, and ultimately the climate risks beginning to affect insured assets. Any loss of income from their investments would be made up by a reduction in claims over the long term. This strategy, he felt, offered the only viable long-term solution for the insurance industry to address its exposure.

However, the response by insurers has instead focused on finding ways to adapt their existing business activities to the threat of climate risks rather than trying to mitigate the drivers behind them (Paterson, 2001; Phelan et al., 2011). This response has centred on trying to find ways to strengthen the inherent weakness in their risk assessments, by supporting actuarial analysis with climate models that attempt to predict future changes to the risk landscape. These future changes include how flood lines will change, how exposure to wildfires may shift or how erosion will begin to affect coastal properties. This has encouraged insurers to manage their own exposure to risk by excluding assets or imposing excesses and deductibles that transfer a degree of exposure back to the policyholder. Despite Leggett's best efforts, examples of insurers engaging in more mitigative forms of risk management have remained largely at the fringe of their business activities (Mills, 2009; Paterson, 2001).

Nonetheless, as the impacts of climate risk have risen, more assets are becoming uninsurable. This has increased the vulnerability of both the insurance industry – through a loss of business – and the communities within which they operate, through a loss of resilience.

Understanding climate risk

Concerns over the impact of climate risks to their clients' assets led Santam, a South African short-term insurer, to initiate a project that explored the impact of climate risk for the insurance industry in the local environment in which it operated. Founded in 1918 and with premiums exceeding R9 billion ($1 billion) – almost 20 per cent of South Africa's total insurance market – the company is Africa's oldest and largest short-term insurer (Santam, 2011). Its business activities stretch the length and breadth of the continent, offering numerous lines of underwriting that now include personal and commercial insurance, alongside more specialised lines such as marine, crop and heavy industry.

The Eden Study had two primary aims: first, to better understand the physical drivers behind climate related risk, and second, to understand how insurers have traditionally understood and responded to these risks. An interdisciplinary partnership of insurers and climate, ecological and social scientists (including the authors of this paper) was formed to explore these questions.

The Eden Study

The Study was undertaken in the Eden Municipality, a coastal region on the Southern Cape coast of South Africa, equidistant between Cape Town and Port Elizabeth (see Figure 8.2). The area was selected as a pilot site because it has borne the brunt of a number of extreme and volatile weather events that have led to significant economic losses for the local community, the government and the insurance sector, over the past decade.

Between 2003 and 2008, a series of floods, droughts, storm surges and wild fires cost the region an estimated R2.5 billion ($300 million) (Holloway et al., 2010). Although 70 per cent of the losses were borne by the South African government, in the form of disaster relief and reconstruction, Eden still accounted for almost 80 per cent of South Africa's entire 'special peril' (or catastrophic) losses for Santam over this period (Nel et al., 2011).

The study explored the impact of three sets of climate-risks in the Eden District: wild fires, flooding and storm surges. Climate modelling, used as part of the study, demonstrated what an expected temperature

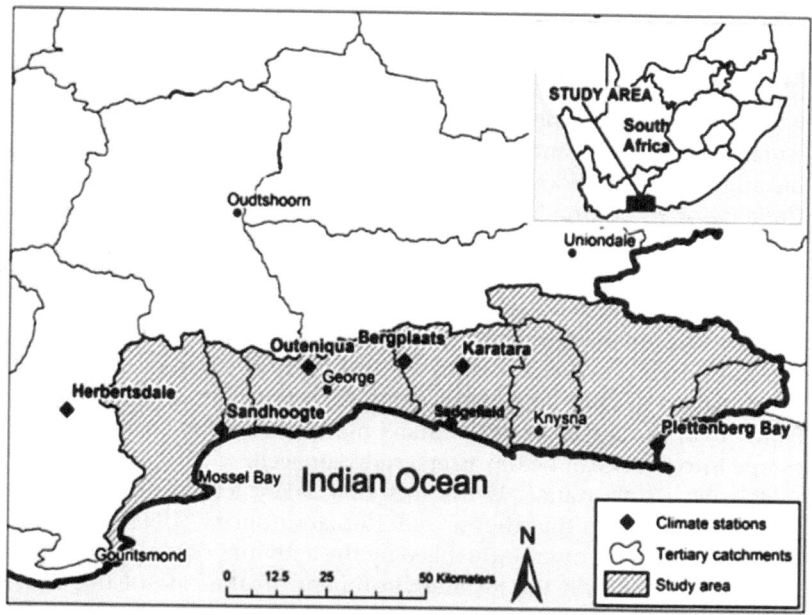

Figure 8.2 Map of the Eden Municipality, South Africa

rise of 1° C by 2040 might mean for Eden in terms of exposure to extreme weather events. They predicted that between 2020 and 2050, Eden is likely to see an increase of 10 per cent in the frequency of intense rainfall days (i.e. >20 mm) and 41 per cent in high fire-risk periods, while the frequency of extreme storm surges will rise six-fold (UNEP-FI, 2011). These climatic changes, if they transpire, are likely to lead to further losses for both the local economy and the insurance sector, in the shape of floods, wild fires and coastal erosion.

In addition, the modelling also highlighted the crucial distinction between these changing local climate systems (which the study refers to as the 'ultimate drivers') and other features within the physical landscape (the 'proximate drivers') that shaped hazards within the local environment. Significantly, it was found that these proximate drivers have 'the same if not greater effect' than climatic drivers in shaping the climate-risks to which Eden is exposed (Nel, 2011). The impact of flooding, for instance, is as much a product of changes to the physical landscape (such as land-surface hardening, deforestation or the management of storm-water drains) as changes in local rainfall patterns linked to climate change.

climate risk = ultimate drivers + proximate drivers

One illustrative example is that of a small municipality within Eden, sandwiched between the Indian Ocean to the south and a large, natural estuary to the north. Since 2003, floods have repeatedly affected the area, causing significant flood losses for the local economy and insurance. Flooding often occurred following periods of intense rainfall, known as 'cut-off lows', that swelled the estuary beyond capacity, causing it to overflow and inundate surrounding communities. While increasingly intense rainfall (the ultimate driver) was an important feature of these floods, a number of proximate drivers were also identified. These included local planning decisions that had allowed development to occur in high-risk flood plains – poorly timed clear-felling of forestry plantations and the failure to artificially breach the estuary mouth in time to enable water to escape into the ocean before assets were damaged.

Of crucial importance, is the fact that a key feature linking these proximate drivers is that they are all 'human-induced' (UNEP-FI, 2011), and their impact is often influenced by the activities of a range of stakeholders active within the local environment. In the case of the above this included the local planning department, the forestry industry and conservation agencies.

This highlighted how, although insurers have had relatively little capacity to mitigate the broader impact of climate change (aside from Leggett's proposal to globally support a transition to lower carbon economies by shifting investment strategies), it does have considerable potential capacity to mitigate local exposure to climate risk significantly, by enrolling the stakeholders who can shape proximate drivers. This offers insurers 'the potential to offset most of the future increases in risk related to climatic changes' (UNEP-FI, 2011). The argument made by the study was that, as a consequence, insurers have more influence than they think they have.

The report concludes with two main recommendations: first, that insurers need to begin to embrace a more 'systems view' of climate risk, to understand how systems respond under different scenarios, and second that they need to begin to explore ways of complementing their established responses with more proactive risk-management activities.

These recommendations have had a significant impact on the company's understanding of climate-risk and its own ability to take a proactive role in mitigating its exposure. As a consequence of this shift in understanding, Santam has initiated a process intended to transform its approach to 'risk management' and to explore the potential 'role of the insurance industry in shaping societal behaviour' in response to climate risk. The hope is that this will 'eventually impact certain decisions' Santam makes around their 'underwriting and risk exposure' practices (Kirk, 2012: 4).

Santam has not been alone in seeking to shift its response to climate-related risks. For example, the CEO of the Insurance Australia Group (IAG) recently affirmed the need for insurance to 'act on multiple levels' while remaining 'at the forefront of raising awareness and driving action on climate change' (Wilkins, 2010). Indeed, for the industry to even remain viable in the future, he claimed, it must begin to tackle 'the challenges of climate change while balancing the needs of ... shareholders, customers, people, communities and environment' (Wilkins, 2010).

This thinking is also reflected more broadly in a recent statement issued by ClimateWise, an industry association representing 39 global insurers on issues linked to climate change, in a report that argues that insurance 'is one of a broad scope of risk management approaches that can facilitate adaptation to climate change and shore up sustainable development' (ClimateWise, 2010).

Limiting the response to climate risk

Addressing the proximate drivers of climate risk, as proposed by the Eden Study, will require the insurance industry to review how it currently manages its exposure to risk, a response still dominated by a focus on risk assessment rather than risk management. Changing this raises a number of challenges.

Their historical ways of doing

In keeping with Moser and Ekstrom's narrative (see Chapter 1), the insurance industry has shaped its response to climate risk in favour of shorter-term, more adaptive (coping) strategies over longer-term, and thus deeper, transformations of its business activities (2010: 1). This is due to the well-known tendency of institutions, when faced with new challenges, to first exploit established technologies before exploring new ones (March, 1991), as exploitation involves refining their existing practices, while exploration involves the riskier and more organisation-ally challenging option of developing new technologies. However, a focus on exploitation can lead to a narrower range of solutions that could be self-destructive over the long term (March, 1991: 73).

Exploitation is further entrenched by the tendency of institutions – with insurance being no exception – to be shaped by features locked into their organisational structure. The tendency is for solutions to new challenges to be sought first in previously established and proven solutions (Bateson, 1972), including the way they gather informa-tion, embed operational rules and practices, and incentivise certain activities. This approach limits the capacity of organisations to 'think' beyond historically established blind spots (Argyris and Schon, 1978). This is often further entrenched by other institutional features, such as management scorecards, that are usually unaligned with support-ing innovative thinking (Kaplan and Norton, 1992, 2001). And this in turn often leads to a further lack of buy-in by other areas of a company (Dibrel et al., 2011).

In the case of the insurance industry, risk assessment based on actu-arial analysis has proved to be a highly successful and reliable tool with a strong record of profitability in the past. Indeed, traditional insurers' competitive advantage, particularly over newer entrants into the mar-ket – such as banks and direct insurers – is still perceived to lie in the extent of their actuarial data sets. These provide them with the ability to assess more complicated risks due to their longer, and thus more reli-able, claims history.

Therefore, rather than reducing the role actuarial analysis plays, and focusing instead on managing climate risk, insurers have responded by trying to find ways that can help improve actuarial reliability. This explains why they have developed climate-risk models to try and predict how, for example, future flood lines are likely to shift or the spread of invasive alien vegetation will influence the prevalence of wildfires.

This response appears to be a sensible and calculated one based on competitive strategies. Indeed, there is little doubt that risk assessment will always remain a crucial feature of insurers' overall response to managing risk exposure. But in light of the recent growth in climate related losses, focusing on risk assessment alone may simply entrench the prioritisation of shorter-term adaptation over deeper and longer-term strategies that can help actively reduce the impact of climate risk on the assets they insure.

Short-termism

Since the 1980s, the global insurance industry has undergone a shift from mutuals (i.e. being owned by policyholders) to being publicly listed. While demutualisation is largely believed to have been a necessary step to help fund expansion in the face of increasing competition, it brought with it a series of unintended consequences. The most significant was the introduction of shareholders and the need to return regular and consistent short-term profits. This refocused management on satisfying their shareholders' needs by maintaining value, and often at the expense of their longer-term risk-management strategies (Heimer, 1985, 2003).

This pressure for short-term profit has led insurers to engage more in the adverse selection of the risks they are willing to underwrite, intensifying the competition to retain the good assets, while avoiding the riskiest ones (Baker, 2003). However, this competition has also limited insurers' ability to achieve their sustainability goals (Ericson et al., 2000) and adapt to climate risk, as they are now less flexible, or less willing to engage in risk management.

A focus on managing risk over the short term has also led to an increased focus on investment returns, which now often contribute as much to insurers' bottom line as their underwriting profits (Heimer, 1985). Investment portfolios are now crucial in generating additional shareholder value. This may explain why insurers have been so reluctant to extend their climate risk-management strategies into their investment portfolios, as proposed by Leggett (1993), as the threat of losing short-term profits outweighs any benefits linked to a reduction in claims over the long term.

The trend towards short-termism has been further entrenched by the emergence of many new organisations – especially the banks and direct insurers – now offering insurance products. This undermines the ability of traditional insurers to unite around common risk-management objectives as the industry has now become so diversified. This further supports insurers' preference for yearly policies, which they can rapidly re-price, rather than commit to longer-term, and therefore more uncertain (and expensive), policies over which they have less control (Maynard and Ranger, 2011).

Another limitation is the fear that attempting to manage the proximate drivers of climate risk in the broader industry may allow competitors to free-ride off any reductions in risk exposure achieved. This is a significant deterrent, as insurers would then have to account for this to both their boards and shareholders. Furthermore, there is the possibility that if exposure to risk is reduced, competition pressures are likely to lead to lower premiums. This may negate any positive benefits gained in managing risk. This is a classic case of the 'tragedy of the commons' (Hardin, 1968). Engaging the broader risk landscape will benefit other companies, even those not contributing to the collective efforts. To counter this, insurers may initially need to frame their risk management activities as not only seeking to improve underwriting returns but as creating marketing opportunities, supporting their relationships with local regulators or simply maximising the benefits associated with being a market leader.

Insurers are also only likely to reduce their focus on short-term profits if they respond as an industry. There are some examples of this in the past. In the 1980s, following efforts by Ralph Nader to make seatbelt use mandatory in vehicles in the United States in order to reduce fatalities, the insurance industry took the National Highway Traffic Safety Administration (NHTSA) to court, successfully demanding that legislation be drafted making the introduction of seatbelts mandatory in all new motor vehicles (Nader, 1987; Gantz and Henkle, 2002).

More recently, there have been attempts to unite the industry around climate collaboration via the development of sustainability principles. These include the ClimateWise Principles (ClimateWise, 2007) and the United Nations Environmental Programme – Finance Initiative (UNEP-FI) 'Principles for Sustainable Insurance' (UNEP-FI, 2007). Yet, while both are steadily growing in support, insurers' response to these initiatives appears marked more by broad and vocal interest, than actionable policy change on the ground.

However, the strongest support is likely to come from national associations representing the broader industry, thereby helping eliminate the free-rider issues. The Association of British Insurers (ABI), for example, successfully lobbied the UK government over improved flood defences in the North of England in 2000 (Mills, 2009; Herweijer et al., 2009). They threatened to withdraw flood cover altogether should government not agree to invest substantially in flood protection. The government, aware that every £1 ($1.6) spent on flood defences, could save £8 ($13) in flood losses, and therefore wary of becoming the insurer of last resort, agreed (ABI, 2010).

The regulatory environment

The regulatory environment, along with the capacity of the state to both legislate and enforce regulatory rules, can be both an advantage and a disadvantage for insurers in supporting the management of climate risk.

On the one hand, the regulatory environment often acts as an effective shaper of risk, with the potential to enrol other stakeholders. Government development strategies, regulation of minimum standards (such as in the construction industry), enforcement of land-use planning policies or even the way storm-water systems are engineered and managed, can significantly influence the impact climate risks have on insured assets.

However, much of the recent trend in regulation suggests that governments also remain trapped into responding to shorter-term challenges, rather than addressing risks more systemically. Some of this emerging regulation can be linked to poor housekeeping by insurers (Grace and Klein, 2009), and is focused on controlling how insurers issue policies, settle claims and manage their risk appetite and investment portfolios. This suggests that governments have yet to appreciate fully the crucial role the insurance industry can play in climate risk governance (Paterson, 1999). This has led to a tendency for insurers to focus more on satisfying their regulatory obligations than on developing a more coordinated response to managing climate risk (Hecht, 2008). This, in turn, has disincentivised the insurance industry from exploring alternative solutions to managing climate risk (Haufler, 2006).

There is also a trend, particularly in more developed markets, of regulation becoming increasingly politicised around partisan interests. The increased cost of insurance following the 1992 Hurricane Andrew in the US, for example, preceded a shift in where insurers were willing to underwrite. This led to a wave of negative public reaction that sharply politicised the regulation of insurance in Florida, and it became a key feature of the 2000 presidential election campaigns (Herweijer et al., 2009). Insurers were obliged to offer policies at depreciated premiums, unrelated to actual exposure, leading to a higher rate of insolvencies following Hurricane

Katrina in 2005 (Grace and Klein, 2009). More recently, in Australia, the eight official natural disasters that occurred in 2011, totalling $5 billion in losses, has led to a move by the government to consider more stringent regulation of insurance due to widespread public frustration over unpaid or unresolved claims (Booth and Williams, 2012).

In areas where the regulatory environment is overly weak, and the state is unable to fulfil its mandate as *the insurer of last resort*, excessive regulation of the insurance industry may become a strategy for governments seeking to transfer their own exposure to climate risk to others. But while a weak regulatory environment can limit the availability of tools for managing risk, it can also prompt the emergence of more innovative risk-management solutions from the private sector, as they seek alternative solutions. This suggests that the strength of government regulation alone is not a determining factor in whether insurers can or cannot become more proactive in managing climate risk. Indeed, the Eden Study emphasised that continuing to focus on government regulation alone effectively shifts focus away from the ability of insurers to emerge as regulators in their own right. In areas that lack a 'shadow of hierarchy' (Börzel and Risse, 2010), government may not even be the most effective regulator, but instead other more influential stakeholders, such as the insurance industry, may take the lead.

Conclusion – Towards climate risk management

The insurance industry's long history as our 'lightning rod' (Mills, 2005) – absorbing many of the shocks that society faces – has firmly established it as a key pillar underpinning economic growth and stability. Yet while the industry's approach to managing risk, predominantly through risk-assessment has worked well in the past, a recent increase in losses, associated with climate change, is throwing this into question. Of greatest concern is the growing realisation that the past is no longer as indicative of the future as it once was, raising fundamental questions about the viability of actuarial analysis.

Insurers have responded to this by investing in climate models to help improve their ability to assess and exclude risk and thus protect both themselves and those assets left in their risk pools. But even these models are plagued with high margins of error, suggesting that insurers must move away from one of their most defining assumptions – that nature is stable and therefore predictable.

Leggett and others attempted to frame their hopes on what was needed to promote a more sustainable, low carbon economy of the future. They turned to the insurance industry as a sector fundamentally threatened by climate change and with the influence to shape economic development at global scales. Yet what they failed to consider was the continuing influence of the existing technologies and practices that had come to define insurance.

What the analysis of this chapter reveals is that a response to climate risk is most likely to occur when it fits within the existing framework of insurers risk-management practices. More specifically, it argues that this is most likely to occur, in relation to climate risk, if the response is focused at the local rather than global scale, due to the faster *return on investment* that can be achieved by engaging closer to the point at which climate-risk impacts insured assets. However, an important limitation of this study is that it is restricted to a local short-term insurer. It is likely that one may find a very different set of reactions with other stakeholders within the industry, such as reinsurance.

Insurance is one of the most enduring institutions we have. Its ability to adapt to new forms of risk and new operating environments throughout its long and illustrious history (Lobo-Guerrero, 2008) is largely unparalleled. Although climate change presents risks the industry has never before seen or imagined, insurance remains the one institution with the clear capacity, and mandate, to adapt itself and the societies within which it operates to the myriad challenges likely to be faced over the coming decades.

Note

We would like to thank Vanessa Otto-Mentz for her support and advice in the preparation of this chapter.

1. This work is based on research supported in part by the National Research Foundation (NRF) of South Africa. Opinions, findings and conclusions or recommendations expressed herein are those of the authors, and the NRF accepts no liability whatsoever in this regard.

References

Argyris, C. and Schön, D.A. (1978) *Organizational Learning: A Theory of Action Perspective* (Reading: MA Addison-Wesley).

Association of British Insurers (2010) Fighting Flood Risk Together. London: Association of British Insurers.

Baker, T. (2003) 'Adverse Selection and Risk Classification', in R. Ericson and A. Doyle (eds) *Risk and morality* (Toronto: University of Toronto Press).

Bateson, G. (1972) 'Steps to an Ecology of Mind: Collected Essays in Anthropology, Psychiatry', *Evolution and Epistemology* (New Jersey and London: Jason Aronson).

Booth, K. and Williams, S. (2012) 'Is Insurance an Under-Utilised Mechanism in Climate Change Adaptation? The Case of Bushfire Management in Tasmania', *Australian Journal of Emergency Management* 27, 38–45.

Börzel, T.A. and Risse, T. (2010) 'Governance without a State: Can it Work?', *Regulation and Governance* 4, 113–134.

Business Week (2011) Insurers' 2011 Disaster Losses Worst Since 2005, Swiss Re Says. http://news.businessweek.com/article.asp?documentKey=1376-LW7 CT46KLVRE01-6AC7D8NBN3STTILC19IGKRTGEJ, date [accessed 17 February 2012].

Cole, D. and Grossman, P. (1999) 'When is Command and Control Efficient?', *Wisconsin Law Review* 887–938.

ClimateWise (2007) The ClimateWise Principles. http://www.climatewise.org.uk/storage/The%20ClimateWise%20Principles.pdf, date [accessed 1 March 2010].

ClimateWise (2010) Global insurance industry statement on Adapting to climate change in developing countries (London: ClimateWise).

Dibrell, C., Craig, J.B. and Hansen, E.N. (2011) 'How Managerial Attitudes Toward the Natural Environment Affect Market Orientation and Innovation', *Journal of Business Research* 64, 4, 401–407.

Durant, R.F., Chun, Y., Kim, B. and Lee, S. (2004) 'Toward a New Governance Paradigm for Environmental and Natural Resources Management in the 21st Century?', *Administration Society* 35, 6, 643–682.

Ericson, R., Barry, D. and Doyle, A. (2000) 'The Moral Hazards of Neoliberalism: Lessons from the Private Insurance Industry', *Economy and Society* 29 4, 523–558.

Farrier, D. and Whelan, R. (2004) '(Why) Do We Need Threatened Species Legislation?', in P. Hutchings, D. Lunney and C. Dickman (eds) *Threatened Species Legislation Is It Just An Act?* (Royal Zoological Society of NSW).

Freeman, J. and Farber, D.A. (2005) 'Modular Environmental Regulation', *Duke Law Journal* 54, 795–909.

Gantz, T. and Henkle, G. (2002) Prevention Institute: Seatbelts: Current Issues at: http://thrive.preventioninstitute.org/traffic_seatbelt.html.

Gelbspan, R. (1998) *The Heat is On: The High Stakes Battle over Earth's Threatened Climate* (New York: Perseus).

Grace, M. and Klein, R. (2009) 'The Perfect Storm: Hurricanes, Insurance, and Regulation', *Risk Management and Insurance Review* 12, 1, 81–124.

Gunningham, N., Grabowsky, P. and Sinclair, D. (1998) *Smart Regulation: Designing Environmental Policy* (Oxford: Clarendon Press).

Gunningham, N. and Sinclair, D. (2002) *Leaders and Laggards: Next Generation Environmental Regulation* (Sheffield: Greenleaf).

Hardin, G. (1968) 'The Tragedy of the Commons', *Science* 162, 1243–1248.

Haufler, V. (2006) 'Insurance and Reinsurance in a Changing Climate', *The Encyclopedia of Earth* 15.

Hecht, S.B. (2008) 'Climate Change and the Transformation of Risk: Insurance Matters', *UCLA Law Review* 55, 1559–1620.

Heimer, C.A. (1985) 'Reactive Risk and Rational Action. Managing Moral Hazard', in *Insurance Contracts* (Berkeley, Los Angeles and London: University of California Press).

Heimer, C.A. (2003) 'Insurers as Moral Actors', in R. Ericson and A. Doyle (eds) *Risk and morality* (Toronto: University of Toronto Press).

Herweijer, C., Ranger, N. and Ward, R. (2009) 'Adaptation to Climate Change: Threats and Opportunities for the Insurance Industry', *The Geneva Papers on Risk and Insurance: Issues and Practice* 34, 360–380.

Holley, C., Gunningham, N. and Shearing, C. (2011) *The New Environmental Governance* (Oxford: Earthscan).

Holloway, A., Fortune, G. and Chasi, V. (2010) *Risk and Development Annual Review* (RADAR): Western Cape, 2010. Rondebosch: Disaster Mitigation for Sustainable Livelihoods Programme, University of Cape Town.

Intergovernmental Panel on Climate Change (IPCC) (2007) 'Summary for Policymakers', in B. Metz, O.R., Davidson, P.R. Bosch, R. Dave and L.A. Meyer (eds) *Climate Change 2007: Mitigation. Contribution of Working Group III to the Fourth Assessment Report of the Intergovernmental Panel on Climate Change* (Cambridge: Cambridge University Press).

Kaplan, R.S. and Norton, D.P. (1992) *The Balanced Scorecard – Measures that Drive Performance*, in E.M. Krekel, C. Diensberg, and B. Schobert (eds) *Harvard Business Review* 70, 1, 71–79.

Kirk, I. (2012) The ClimateWise 2012 Thought Leadership Series: Issue Two: The Value of Ecosystem Resilience to Insurers (London: ClimateWise).

Leggett, J. (1993) *Climate Change and the Insurance Industry: Solidarity among the Risk Community* (Amsterdam: Greenpeace International).

Lobo-Guerrero, L. (2008) 'Pirates, Stewards, and the Securitization of Global Circulation', *International Political Sociology* 2, 219–235.

March, J. (1991) 'Exploration and Exploitation in Organizational Learning', *Organization Science* 2, 1, 71–87.

March, J. and Simon, H. (1958) *Organizations* (New York: Wiley).

Maslakovic, M. (2011) *Financial Market Series: Fund Management* (London: The City UK).

Maynard, T. and Ranger, N. (2011) 'What Role for "Long-Term" Insurance in Adaptation? An Analysis of the Prospects for and Pricing of Multi-Year Insurance Contracts', *Change* 26.

Millenium Ecosystem Assessment (2005) *Ecosystems and Human Well-Being: Health Synthesis* (Geneva: World Health Organisation).

Mills, E. (2005) 'Insurance in a Climate of Change', *Science* 309, 1040–1044.

Mills, E. (2009) *From Risk to Opportunity: Insurer Responses to Climate Change* (Berkeley: Ceres).

Moser, S.C. and Ekstrom, J.A. (2010) 'A Framework to Diagnose Barriers to Climate Change Adaptation', *Proceedings of the National Academy of Sciences of the United States of America* 107, 51, 22026–22031.

Munich, R.E. (2012) *Worldwide Natural Disasters 1980–2011*: Overall and Insured Losses Munich RE.

Nader, R. (1987) 'Loss Prevention and the Insurance Function', *Suffolk University Law Review* 21, 3, 679–689.

Nel, J., Le Maitre, D., Forsyth, G., Theron, A. and Archibald, S. (2011) *Understanding the Implications of Global Change for the Insurance Industry: The Eden Case Study* (Stellenbosch: Council for Scientific and Industrial Research).

Paterson, M. (1999) 'Global Finance and Environmental Politics Climate Change', *Ids Bulletin*, 30, 3, 25–30.

Paterson, M. (2001) 'Risky Business: Insurance Companies in Global Warming Politics', *Global Environmental Politics* 1, 4, 18–42.

Phelan, L., Taplin, R., Hendersen-Sellers, A. and Albrecht, G. (2011) 'Ecological Viability or Liability? Insurance System Responses to Climate-risk', *Environmental Policy and Governance* 21, 112–130.

Porter, M.E. and Kramer, M.R. (2002) 'The Competitive Advantage of Corporate Philanthropy', *Harvard Business Review* 80, 56–68.

Santam (2011) Santam: Annual Report 2011.

Stern, N.H. (2007) *The Economics of Climate Change: The Stern Review* (Cambridge: Cambridge University Press).

The Geneva Association (2011) Global Insurance Industry Fact-sheet (Geneva: The Geneva Association).

UNEP FI (2007) Insuring for Sustainability. Why and How the Leaders are doing it www.unepfi.org/publications/insurance/ date [accessed 20 June 2011].

UNEP-Fi (2011) *Insurance in a Changing Risk Landscape: Local lessons from the Southern Cape of South Africa*, Geneva.

Wilkins, M. (2010) 'The Need for a Multi-Level Approach to Climate Change – An Australian Insurance Perspective', *The Geneva Papers on Risk and Insurance Issues and Practice* 35, 336–348.

9
Of Culture and Religion: Insurance Regulation and the Informal Economy in a South African City

Moliehi Shale

Introduction

In many emerging economies with areas of limited statehood, there are numerous obstacles to effective governance in response to complex challenges, such as those associated with the impacts of climate change. In the city of Cape Town's informal settlement areas, flooding is a frequent occurrence and is regarded as a climate-induced threat. The Cape Town metropolitan area is one of South Africa's oldest and fastest-growing urban regions. Langa and Philippi townships are located on the Cape Flats (see location map in Figure 9.1). This research study explores the flood response strategies of small-business owners through burial societies in these two areas. Burial societies are an informal self-insurance mechanism based on voluntary commitment and reciprocity (ECIAfrica Consulting, 2003).

At the beginning of democratic government rule in 1994, neo-liberal policies replaced apartheid policies in South Africa. The Growth, Employment and Redistribution strategy (GEAR) was introduced by the South African government in 1996 and prescribed 'more freedom to the market, less regulation of international trade and more integration into the global economy' (Miraftab, 2004). GEAR's aim was to promote growth in atypical areas of employment, reduce firms' costs and increase investment (Valodia, 2001). However, contrary to its intended outcomes, 'the state's neo-liberal policies on trade, industry and the labour market have opened up the industrial sector to international competition – leading the economy into one of capital intensification' (Valodia, 2001: 881). This has been coupled with extensive job losses and has reinforced the divide between the access that different population groups have to basic urban services (Miraftab, 2004). Neo-liberal

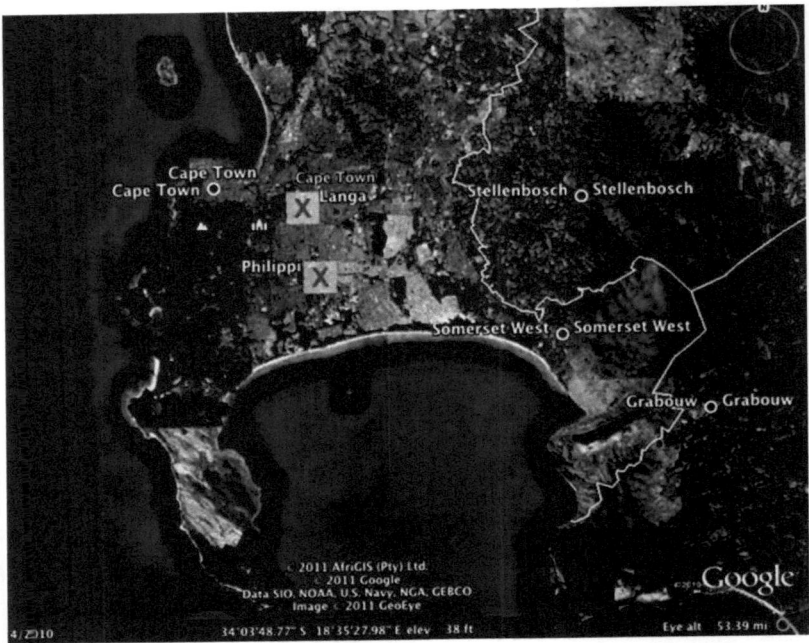

Figure 9.1   Langa and Philippi townships, Cape Town, South Africa
Source: Google Earth (2010).

policies have also pushed many workers into the informal sector where labour protections are limited, competition is high and state-led service provision is poor (Lindell, 2010; Hansen and Vaa, 2004).

Informal settlement areas are home to many of South Africa's informal businesses, and they are a characteristic of limited statehood because they point to weak land-use policy regulation and enforcement. The city of Cape Town's municipal government has responded to the growing migration into informal settlement areas by relocating people – often further away from the city centre. Attempts to restrict migration into these environmentally vulnerable areas have been unsuccessful due to weak monitoring of the expansion of the settlements (Ziervogel and Smit, 2009; Wood, 2004). Meanwhile, the number of informal businesses has grown in keeping with the increase in the numbers of people in these settlements.[1]

This chapter examines the use of civic self-help governance as a form of climate change risk insurance for small-business owners. Burial societies are specifically examined as civic organisations because their prevalence and use has grown with the growth of urban constituencies in South Africa, where they provide burial insurance, savings and credit facilities. The chapter emphasises the importance of this type of insurance at an institutional level, and I discuss how research on this topic has been tackled through examining the literature. I will develop a theoretical explanation for, and provide an empirical exemplification of, how and why burial societies are so important in the risk-management strategies of their users. The study will show the role that their regulations play in building networks of trust that users can draw on to respond to climate-induced hazards, particularly flooding.

This chapter examines qualitatively the circumstances that lead poor urban-based small-scale business owners to take out insurance through burial societies rather than formal insurance companies offering flood insurance. Poor constituencies have little incentive to seek flood protection from the state, which is in any case virtually absent, or to use formal sector insurance whose operations use rules very different to those in the informal business sector. Further, the networks built through burial societies contribute to flood management because the societies' activities and regulations expand beyond the boundaries of burial society regulation.

In this study, the unit of analysis is business owners operating small-scale, unregistered enterprises in informal settlement areas. These types of informal enterprise are a common feature of the urban African economic landscape – providing key services such as employment, shelter, transportation and various other social services (Stren and Halfani, 2001). Here, the informal sector is understood as the extra-legal and unregistered activities that constitute the informal city (Hansen and Vaa, 2004) and that circumvent state regulation (Lindell, 2010; Castells and Portes, 1989). However, being outside the regulation of the state does not mean that they are completely unregulated. Lindell (2010) points to the variety of actors and institutions (the institutional field) in the informal economy through which rules are made beyond the state, a matter to which I will return below.

This chapter adopts Scott et al.'s (2000) definition of institutional structure, according to which the institutional field is composed of three pillars (cognitive, normative and regulatory) that exist simultaneously in an institutional field, albeit with varying degrees of influence. The informal business sector can give rise to particular informal governance institutions. For example, in burial societies the cognitive structures through which a shared idea of the social environment is constructed are largely based on religious and cultural beliefs and norms around death, such as prayer for the dead. The normative framework consists of values and beliefs that are largely centred on enforcing mutual obligations between actors – encouraging relations of reciprocity. Lastly, the regulatory system recognises the authority of the institution to oversee conformity and even to impose sanctions where there is a lack of conformity. Sanctions such as fines are very common in burial societies and commonly upheld by a society constitution.

In different areas (as opposed to whole territories), informal governance institutions often emerge to provide social and political order as well as collective goods where 'formal state institutions have ceased to exist or to provide governance services' (Risse, 2011: 14). Rather than acting against state institutions or in the complete absence of political order, 'the various institutionalised modes of social coordination produce and implement collective binding rules, or provide collective goods' (Risse, 2011: 9).

Networked and decentred conceptions of governance have gained much traction in the governance literature in its effort to better appreciate the complexity of social organisation and the creation of governance systems that complement or reflect how the systems function (Burris et al., 2005). Data from a case study in the Victoria Mxenge informal settlement in Philippi Township and the Joe Slovo informal settlement in Langa Township, Cape Town, will help to illustrate these assertions.

The two case studies are used to reveal the challenge to mainstream governance literature in explaining how the urban poor protect their businesses from climate-induced hazards, without using either market-based insurance instruments or state-led safety nets. The study starts with an interesting puzzle. The sites studied here flood annually, which causes enormous damage to businesses and homes. Yet, despite the availability of cheap insurance policies[2] in the formal market, small-business owners here use their burial society membership as a means to insure themselves against climate-related hazards. Unlike formal insurance, burial societies do not offer financial compensation. Instead, they ensure access to culturally significant services and practices such as an acceptable funeral and access to equipment and services – including in-kind services such as cooking and prayer – to their members (ECIAfrica Consulting, 2003).

Climate change as it affects poor communities through flooding

Among the main causes of vulnerability in poor communities are the fragile environments in which they live in and the lack of state-led responses. In addition, many poor communities operate in fragile economies whose resources often cannot be used effectively to respond to recurring climate-induced hazards such as flooding.

Many poor people do not qualify for formal insurance in the first place, because they do not have any formal employment and cannot afford to travel to urban centres where work is available. Many poor people seek services in the informal sector, which may further deter them from approaching the formal sector for insurance (Matul et al., 2010).

Theoretical orientation

Governance theory, as initially formulated, suggests that the state provides all communal resources (Wood and Shearing, 2007). Public sector safety net transfers are commonly targeted at the poor, to prevent them from falling even further into poverty. One form of safety net transfer is insurance.

Both the adaptation and resilience literatures identify insurance-related financial arrangements as some of the key regulatory institutions in the management of climate-induced hazards in developing economies (IPCC, 2012). However, 'our understanding of the role of insurance tools in helping poor constituencies to adapt to climate change remains limited' (Bals et al., 2010: 1). Unlike developed economies, providing state-led and formal insurance to the poor has proven to be an enormous challenge in many developing economies (Toman, 2010). As a result, poor households seek insurance services in the informal sector (Murdoch, 1999).

One of the reasons for low formal insurance uptake among the poor is affordability. Another hurdle is knowledge of insurance and its use as a safety net against the effects of flooding. Bals et al.'s (2010) Indian example points to high insurance uptake after a major climate-induced catastrophe, but the insurance is usually for life and assets and are short lived. In democratic South Africa, the law has given significant attention to environmental issues, which is reflected in the country's environmental legislation. Many of these policies, including the Constitution, require that the government provide basic services such as food, shelter and education in a safe environment. These governance policies have earned the country international recognition for addressing issues of environmental justice (Patel, 2009).

Despite their world acclaim, South African environmental policies have been ineffective at ensuring that the poor and marginalised live in a safe environment, and this can be attributed to the complex nature of the environmental problems. For example, climate change governance concerns are an area of great disagreement in South Africa, given the problems of limited statehood and mounting social problems of the nation (Madzwamuse, 2010).

At the local level, the impact of climate change is particularly challenging in the resource-deprived communities that are most vulnerable to impacts such as flooding. The communities may order their choices in a manner that prioritises other needs, such as food, clothing and shelter, and may relegate environmental needs to the bottom of the list (Darkey, 2012). Also, those most vulnerable to flooding may not have adequate representation or access to higher levels of governance and decision-making, and they may be unable to articulate possible solutions to help them adapt to the severe threats they face (Ostrom, 2010; Agrawal, 2008). Consequently, environmental management is often seen as a problem to be resolved by the government. For example, in his study on the attitudes and perceptions of female heads of household in Mamelodi Township, Darkey (2012) found that they prioritised socio-economic issues such as education, job opportunities and the provision of services over environmental issues. The author concluded that addressing these socio-economic concerns is key in getting people more interested in environmental issues.

Municipal- and national-level responses to climate impacts in South Africa have also had major shortcomings, particularly as a result of weak policy regulation and enforcement (Mokwena, 2009; Mukeibir and Ziervogel, 2007). The literature does not fully explain why the urban poor do not make more use of formal insurance (despite its availability to low income earners) but it does suggest that work and wellbeing in these environments is supported by the norm-based institutions embedded in the informal economy (Casson et al., 2010).

Methods

The chapter presents two empirical case studies in two South African townships in the city of Cape Town: Victoria Mxenge in Philippi and Joe Slovo in Langa Township (See Figure 9.1).

The selection of the research sites is based on their vulnerability to flooding and the presence of a visible informal sector. The main difference between the two research sites is the period over which they have been settled. Langa is the oldest township in Cape Town and was established in 1927. Philippi developed in the 1980s.

The particular focus of my research on burial societies was informed by the initial data collected in the study. This showed a high rate of membership in burial and savings societies, although there was no insurance specifically for the effects of floods. The low uptake of formal insurance was especially surprising in light of the recurrence of floods in both areas and the availability of formal insurance products targeted at low-income earners. To understand the function of burial societies, primary data was collected through interviews with small-business owners, burial society members (and non-members) and other stake-holders, such as burial society committee members. The data collection particularly sought to understand the contribution of burial societies in insuring the businesses of small-scale small-business owners against climate-induced hazard and resultant business losses.

The initial research respondents were selected using a convenience sampling method, and the snowball method (Babbie and Mouton, 2001) was later used to identify and recruit further respondents. The sample of 37 respondents was made up of former and current small-business owners in order to capture the ways in which they (successfully and less successfully) responded to flooding in relation to their businesses.

Civic self-help governance in South Africa

The literature on the role of institutions in economic activity has gen-erally been understood in terms of the formal rules and regulations governing economic activity, including tax laws, state-led regulations and infrastructure, and the impact of these on the economic sector (Casson et al., 2010). 'In the absence of welfare states' protective social security, new forms of self-organisation and civil self-help governance have developed where the state has withdrawn' (Jürgens, 2012: 154). Economic returns and informal survival strategies have taken the place of state-directed development initiatives including social security serv-ices such as insurance.

Civic self-help associational activity all over the world is reflected in private voluntary organisations whose activities pursue public objectives outside the realm of the state (Robinson and White, 1997). Robinson and White point to three sets of pressures that have led to global growth in civic activity: escaping political oppression; external assistance that has boosted the resources available to grassroots organisations; and gov-ernments' privatisation of public services, thereby increasing the role of churches and non-governmental associations in national development programmes.

Civic associations have a long history in South Africa, particularly in the politics of revolt against the apartheid state. Civic associations in South Africa also tend to be formed along religious and cultural principles, and in this sense, they are carriers of positive values and morals required for cohesive communities (Tostensen et al., 2001). Their long history among poor South Africans also shows their popularity in maintaining the personal relations and cultures that sustain the informal market and regulate economic activities. 'This gives socially regulated structures a competitive edge over formalised ones' (Roberts, 1994: 8). White's (1998) analysis of voluntary associations reveals their role in the development of a like-minded culture that facilitates participation and action in areas that address the allocation and management of resources to respond to collective problems (UNDP, 1997) both with and without the participation of government institutions.

Although the characteristics of voluntary associations are often made with reference to the state, they may also be completely neutral vis-à-vis the state (Tostensen et al., 2001). Common examples of civil associations that do not hold a counter-position to the state can be seen in networks of mutual help, such as rotating savings and credit associations (ROSCAs), which tend to grow in low-income social structures. In South Africa, ROSCAs are commonly referred to as stokvels, and they are a long-standing feature in the economic lives of (black) Africans (Porteous, 2003).

ROSCAs are widely used in poor constituencies in developing economies to pool limited resources and spread risk among members. Risk spreading is not unique to ROSCAs. In fact, insurance in the formal economy is based on the very principle of pooling financial resources from members of a network to pay for losses when they occur (Mills, 2004; Anderson and Baland, 2002). Civic self-help associations give an important and relevant theoretical context to the research question because they show the influence of social and religious guidance in these associations. The following discussion of burial societies identifies them as a type of civic self-help association whose networks of mutual help are used in response to flood impacts.

Burial societies: A form of Civic Association

Burial societies are a common type of civic association in poor South African constituencies. The literature paints a picture of burial societies as informal institutions concerned with death and its associated stresses, such as loss of household income and costs of burial (Thomson and Posel, 2002). My research findings show that the benefits of burial societies are far greater than those pointed out by Thomson and Posel (2002). Geertz (1962) called this rotating communalism: 'This typology of insurance often takes the shape of rotating credit associations, which use customary patterns of cooperation, mutual help, and communal responsibility to regulate the emergent activities' (Geertz, 1962: 262). Geertz also found that the reciprocal actions between individuals softened the harshness of the economic aspects and maintained customary ties and traditional moral values.

Bähre's (2007) ethnographic study of self-help groups in Cape Town, South Africa, characterises burial societies as communal efforts, often fulfilling financial as well as cultural roles. Formal insurance products also spread risk among their members, although they employ different regulatory mechanisms. For example, the formal insurance sector requires loss prevention behaviours for its sustainability (Bähre, 2007). Civic associations also have regulatory codes designed for loss prevention but members often do not have any historical credit information. These institutions rely on the testimony of the community of which members are a part, in order to ensure preventative behaviours. For example, stokvels (Thomson and Posel, 2002; Verhoef, 2001; Lukhele, 1990) in South Africa embody important African cultural traits of sharing, support and mutual help (Kritzinger, 1996), and these traits play a useful role in discouraging members from defaulting on payments and encouraging the assistance of others. The principal reasons for the existence of burial societies can be summed up as follows: first, burial societies provide a stable environment through which a key service, primarily death insurance, can be solicited. Second, the rules are socially acceptable as they reaffirm their members' cultural and spiritual needs. Third, burial societies provide a social network that is responsive to the immediate needs of the communities that they serve, including responding to the impacts –such as flooding – of a changing global climate.

Burial societies in poor African communities

Under the apartheid system, workers lived in barracks, where they were often organised according to their place of origin (Bähre, 2007). Burial societies in these communities ensured proper burial in the members' homeland (Thomson and Posel, 2002; Verhoef, 2002; Kritzinger, 1996). Most importantly, burial societies ensured that their users' funerals followed shared cultural norms of the group. For example, Bähre (2007) found that groups of Xhosa migrants in Cape Town relied on burial societies to ensure 'proper' burial in their homeland in the Eastern Cape (Bähre, 2007). A 'proper' burial is typically associated with following precise customs of individual communities, including entertaining friends and family attending the funeral for as long as they choose to stay and assisting the family of the deceased (Case et al., 2008; Roth, 1999), and it marks the strength of ties between the deceased person's family and the surrounding community. While formal insurance products pursue profit above all else (Collair, 1992), burial societies integrate cultural and religious values into their structure (Kritzinger, 1996). The regulation of burial societies warrants particular attention, because it reveals the influence of its key values on its ability to provide multiple services to its users.

Despite the widespread understanding of insurance, there is little understanding in the literature on its utility and application by poor individuals and groups in the face of hazard. An understanding of burial societies as one aspect of the 'popular economy of solidarity' (Kritzinger, 1996: 110), still fails to make sense of their cultural and religious characteristics and how they influence users' actions in the management of climate-related hazards.

Results

I found the exposure to flooding in both townships to be similar – with residents of both describing annual floods – and I did not notice any significant differences in the composition of the informal sector.

The leadership structure in the two sites was very different, however. In Joe Slovo, people supported a community-elected leader who, at the time of the interviews, was standing for a position in the local council elections. In Victoria Mxenge, the local councillor was a locally elected member of the ANC. Because Victoria Mxenge had an ANC councillor, I assumed that Victoria Mxenge would receive more assistance from the municipal government. Instead, residents in both townships had received little or no assistance from the municipal government, and both were more concerned about other service issues. A young man running a shebeen in Victoria Mxenge complained that when jobs were available, the councillor gave preference to followers of her political party, saying, 'there is a challenge of political favouritism; the ANC supporters get more favour on jobs than COPE supporters'.[3]

Today burial societies have grown in group size and economic value. They are still the main source of insurance for many South Africans who cannot participate in the formal economy. The research data reveal that burial societies operate as businesses and not just as a self-help community initiative or club. In Victoria Mxenge, the treasurer of a burial society indicated that committee members fulfilled their respective roles voluntarily, making the society more like a social club. She also said that the chairman (who is often also the owner) of the society does not receive a salary, although there were provisions in the monthly budget for his transport and telephone expenses. The chairman also has some discretion on how some of the society funds are used towards committee meetings. She cited instances of meetings that ended late in the evening, whe he authorised the use of money towards renting a private vehicle to take the members home safely.[4] These and other running costs were announced in burial society meetings for transparency but members were allowed to approach the committee when they had questions relating to the finances of the society.

The members' proximity to the committee and access to society information gives them confidence that they will receive financial support when they need it, while the social club component satisfies users' social-support needs. Insurance in the formal sector offers only one of these assurances; and the added benefit of joining a burial society becomes clearer when one looks at its categorisation in terms of profit-making business and support-based social club. Statements such as the following point to these added benefits in burial societies: 'It is not about money. That's why we pray and sing at our meetings. We are just helping because for R5000 you cannot bury the person but it is better than nothing ... for me it is because if I died, I would want someone to sing for me. The policy cannot sing for me'.[5]

Burial society businesses reflect what Ruggie (2004) refers to as a business whose public role is to compensate for a governance failure at the local level.[6] To overcome this governance failure, the regulatory principles of burial societies build networks of friendship and trust between members. As a result, many of the participants indicated that they had sought help from neighbours, friends, family and fellow burial society members in the last flood incident. A young shebeen[7] owner in Philippi recounted a time when the shack[8] that he works from was damaged by water and he got a loan from his uncle: '... the money was repaid with no interest. The money was useful. With the money I bought stock for the business, bought groceries and paid the burial society contribution'.[9]

The trust between members is developed and maintained through the burial society's recruitment process, in which existing members recruit members from their own close social network to regulate the risk of defaulting and other forms of malfeasance, such as fraud. Here, social and cultural characteristics of users are important in creating trust among members (see MacLeod, et al., 1997 and Amin and Thrift, 1992). A burial society committee member indicated that the burial society is big, and is divided among varied constituencies for easy management. In response to the difference in these constituencies, she replied: 'the members are mostly black and some coloured people. The blacks are not all Xhosa speaking but all our members join because they like our service'.[10]

Assurance of monetary and other forms of help was another result of the strict rules of burial societies. For example, one of the reasons for a preference for burial society membership among research participants as opposed to formal insurance was that there is transparency about the finances of the business because financial reports were read out to the society members at the meetings. 'We pay at Nedbank and take the deposit slips to the meeting on Saturday, unless there is a funeral. The treasurer collects the deposit slips and the society treasurer tells us how much we have in the society account'.[11] Transparency around society finances gives members the assurance that the society has sufficient funds for them to receive help when they need it in return for their participation in the group and helping others when in need.

The influence of religious and cultural components in burial was also evident in the research sites. My data revealed that 51 per cent (19) of the total sample of business owners interviewed in the study were members of a religious institution. The influence of religion can also be seen in the burial society meetings, which always begin with prayer. To illustrate, a member of the Zion church burial society said 'we pray for their [fellow society members] businesses to grow and their products to sell well, we pray for our families, sometimes for abusive husbands'.[12]

The religious principles of the burial society provide a level of certainty that in the event of a crisis, such as death or a flood, there will be assistance from the members of the network. This is based on the belief that those who share a certain level of love for God will show good neighbourliness and mutual assistance to one another. This religious component is supported by the African cultural ideology of 'ubuntu'. 'Ubuntu emphasises respect for fellow human beings while stressing the inclusiveness of African communities' (Bähre, 2007: 135–136). An interesting example of burial societies' commitment to mutual assistance is found in a story about a member who had stopped paying her burial society contributions but soon lost a family member. Because the member had diligently attended meetings and the funerals of other society members, the burial society chairperson decided to reward the member's participation by encouraging members to make small donations to the member's family: 'We want to show our members that we love them but the constitution does not allow us to pay for people that have not been paying their contributions'.[13]

Mutual assistance between community and family members was a common flood response. In response to how they had dealt with the impact of previous floods, and about current preparedness, respondents indicated that they relied on family members, friends and colleagues. Interestingly, however, foreign nationals often indicated that they did not have access to this type of support. A Zimbabwean craftsman in Philippi recounted his first flooding experience in Cape Town shortly after arriving in the city: 'The water would destroy all my things. I didn't have a proper shed to protect the produce. I couldn't afford the shed because I was starting at the time, I didn't know anyone. It happened a lot of times, especially in the winter – like last winter; it was a long winter'.[14] In comparison to the local business owners, the respondent's experience suggests that a lack of access to a network of help can heighten one's exposure to flood impacts.

In sum, respondents had not purchased an insurance policy to protect themselves against the financial risk of flooding. Rather, during and in the wake of flooding, they relied on social support networks among which were those built around burial societies.

Discussion: Religious and cultural embeddedness in civil self-help governance

Despite their popularity in voluntary associations, and burial societies in particular, religious and cultural concepts have been criticised for lacking rationality. For example, Ellis (1962, 1973) has claimed that religiousness promotes maladjustments. In contrast, Mitchell (2003) points to the symbolic impacts that often stem from culture and both historic and contemporary aspects of religion and disaster. Religion and spirituality have become useful for humans in making sense of and controlling their physical environment, but this does not translate to sustainable protection from hazard. Despite these criticisms, Liebow (1966) and Granovetter (1985) caution that what may look like irrational behaviour may be sensible given the present situational constraints under which decisions are made.

The literature often points to a high risk of malfeasance in voluntary associations. While respondents in this study acknowledged that it does occur, they also indicated that it had never happened in their societies and that there were network-based checks and balances in place – such as the society recruitment methods – to ensure that it was avoided.

Despite these protections, the actions of both business owners and users may seem irrational to the onlooker. In terms of the member, it seems irrational because there are other affordable insurance products available in the formal market that would provide financial compensation for losses allowing rebuilding the business, a benefit not received from the burial society. In terms of burial society owners (or conveners) their business model may seem unsustainable because the pricing of their product is not market-related. For example, the burial societies sampled in the research charged the same premiums for all their members, irrespective of their vulnerability to income loss or death.

One interpretation for these seemingly irrational actions on the part of the burial society owners and members is that, where state protections are limited and affordable service provision is scant, the appeal of the cultural and spiritual elements of burial societies becomes more significant. The risk of flooding has therefore created an organisational field in which different social principles are drawn into the business to respond to a common threat, which is flooding. Burial societies are particularly interesting social groups because they have garnered so much support from their users over the years, yet they have neither been destroyed by the formal market nor adopted by national development strategies such as municipal flood risk management strategies. For example, 'the city of Cape Town's flood risk management strategy has largely been unsuccessful because of institutional and governance constraints that result in a focus on narrow technical solutions and on the provision of disaster relief' (Ziervogel and Smit, 2009: 2). The City's attempts to reduce flood risk in its informal settlement areas have not all been unsuccessful. For example, Ziervogel and Smit (2009) point to the success of infrastructure interventions in the form of drainage channels, but longer-term proposed solutions such as the relocation of residents have been met with great resistance, perhaps due to a fear of the social fragmentation that can result from relocation (see Oldfield, 2002). Even though these strategies are geared at the protection of informal settlement dwellers, they fail to acknowledge and build on existing civic networks such as burial societies in crafting more effective approaches to climate change. The aim of this chapter is to point to existing initiatives such as burial societies that can be built on to respond to urban flood risk response in the city's informal settlement areas.

Conclusion

This chapter explored how small-scale business owners in informal settlement areas insure against climate change hazard, particularly flooding.

The findings have shown that, at first glance, small-business owners do not seem to take insurance despite high risk, on the one hand, and the availability of affordable insurance policies, on the other. Yet, looking closer, they do insure through burial societies.

The chapter points to the characteristics of burial societies with an emphasis on their religio-cultural principles as key building blocks of the support network that burial society members call upon for in-kind support when confronted with a climate-induced hazard like flooding. The chapter also discussed the seemingly irrational response to flooding by small-business owners by enrolling in burial societies as opposed to seeking insurance cover targeted at responding to climate events in the formal market.[15]

The chapter also highlights the importance of religio-cultural principles, based on which collective decisions are made and followed. The empirical analysis provides insight into how small business responds to the risk of flooding by building networks of users bound by trust and reciprocity. Through these networks, users access help not only in the event of death, but also when faced with the impacts of flooding. Burial societies play an important role in the governance of flooding by bringing their members together and inculcating mutual support in their networks so that they have access to in-kind support in the face of a climate-related hazard.

The discussion has demonstrated how the rules and principles of burial societies promote mutual help that is then used by members to adapt to the impacts of flooding. Demonstrating how this mutual help is propagated shows how burial societies adopt social norms and principles to build networks of help among users and as a response to the impact of flooding.

Herein lies the answer to the question of how burial societies contribute to climate change governance, particularly as it affects their members in informal settlement areas where flooding is a common occurrence. Not only are the rules and main governing principles of mutual assistance geared strongly towards protecting the association and its members but they also facilitate climate governance by building systems of accountability among their members and offering a mutually shared response to a common threat to their livelihoods. Perhaps what is an even greater contribution to climate change governance is that burial societies facilitate a shared understanding of the risk of flooding as well as the response. Even though business owners' response to flood risk may seem passive and perhaps even fatalistic to the onlooker, when viewed in light of burial societies as an agents of

188 Business and Climate Change Governance

mutual association, its appeal to small-business owners becomes more evident.

Notes

1. The natural annual national population growth rate in South Africa is 1.1 per cent, with the average growth rate of cities being 3 per cent. This has resulted in a continuing proliferation of informal settlements in the country's cities and towns as well as a 3.1 per cent increase in year-on-year change in employment in the informal sector (Statistics South Africa[1,2]).
2. The premiums for a funeral at one of South Africa's oldest insurers, Old Mutual, start from R40 (approximately $5.00) per month (see: http://www.oldmutual.co.za/personal/funeral-cover/funeral-plan/funeral-range. aspx).
3. Interview with barbershop owner, 9 June 2010, Philippi.
4. Interview with burial society secretary and member, 18 June 2010, Philippi.
5. Interview with burial society secretary and member, 18 June 2010, Philippi.
6. The literature identifies other motivations for businesses operating in the informal sector, particularly as an escape from the strict regulation of the formal economy (cf. Tissington, 2009, Devey et al. 2003), but that is beyond the scope of this chapter.
7. Unlicensed businesses or private homes selling alcohol.
8. Commonly used to refer to the housing structures built with different types of materials in informal settlement areas. Some respondents in the research referred to these as a 'hocki'.
9. Interview with shebeen owner, 15 June 2010, Philippi.
10. Interview with burial society secretary and burial society member, 18 June 2010, Philippi.
11. Interview with African crafts business, 12 October 2010, Joe Slovo.
12. Interview with food seller and burial society member, 12 October 2010, Joe Slovo.
13. Interview with burial society secretary and burial society member, 18 June 2010, Philippi.
14. Beaded crafts business owner, 9 June 2010, Philippi.
15. Climate Wise recently reported that its group's two members had an 88 per cent compliance with its six principles of managing risks related to climate change, (Veysey 2012).

References

Agrawal, A. (2008) 'The Role of Local Institutions in Adaptation to Climate Change', *Social Dimensions of Climate Change* (Social Development Department, World Bank Washington DC, USA: The World Bank).
Amin, A. and Thrift, N. (1992) 'Neo-Marshallian Nodes in Global Networks', *International Journal of Urban and Regional Research* 16, 4, 571–587.

Anderson, S. and Baland, J.M. (2002) 'The Economics of RoSCAs and Intra-Household Resource Allocation', *The Quarterly Journal of Economics* 117, 3, 963–995. doi:10.1162/003355302760193931.

Babbie, E. and Mouton, J. (2001) *The Practice of Social Research* (Oxford: Oxford University Press).

Bähre, E. (2007) *Money and Violence: Financial self-help groups in a South African township* (Boston: Brill).

Bals, C., Burton, I., Butzengeiger, S., Dlugolecki, S., Gurenko, E., Hoekstra, E., Höppe, P., Kumar, R., Linnerooth-Bayer, J., Mechler, R. and Warner, K. (2010) *Insurance-Related Options for Adaptation to Climate Change. Executive summary* (The Munich Climate Insurance Initiative MCII).

Burris, S., Drahos, P. and Shearing, C. (2005) 'Nodal Governance', *Australian Journal of Legal Philosophy* 30, 30–58.

Case, A., Garrib, A., Menendez, A. and Olgiati, A. (2008) *Paying the Piper: The High Costs of Funerals in South Africa.* NBER Working Paper No. 14456, October.

Casson, M.C., Della Giusta, M. and U. S. Kambhampati (2010) 'Formal and Informal Institutions and Development', *World Development* 38, 137–218.

Castells, M. and Portes, A. (1989) 'World Underneath: The Origins, Dynamics and Effects of the Informal Economy', in A. Portes, M. Castells and L. Benton (eds) *The Informal Economy: Studies in Advanced and Less Developed Countries* (London: John Hopkins Press).

Collair, I. (1992) *A Review of the Stokvels Movement in Some RSA Townships with Reference to Financial Management Techniques Used in Them.* Unpublished essay for B. Comm. (Honours), University of Cape Town, Cape Town.

Darkey, D. (2012) 'Democratization of Environmental Governance: Perceptions and Attitudes of Township Women Towards the Environment', *Urban Forum* 23, 209–219.

Devey, R., Skinner, C. and Valodia, I. (2003) *Informal Economy Employment Data in South Africa: A critical Analysis.* Unpublished paper delivered at TIPS and DPRU FORUM, 8–10 September, Johannesburg.

ECIAfrica Consulting (Pty) Ltd (2003) Burial Societies in South Africa: History, Function and Scope. A report prepared for DGRV, Johannesburg, South Africa.

Ellis, A. (1962) 'The Case against Religion: A Psychotherapists View', *The Independent* 126, 4–5.

Ellis, A. (1973) 'Albert Elis' Rationality Score', *Rational Living* 8, 2, 31.

Geertz, C. (1962) 'The Rotating Credit Association: A "Middle Rung" in Development', *Economic Development and Cultural Change* 10, 3, 241–263.

Granovetter, M. (1985) 'Economic and Social Structure: The Problem of Embeddedness', *American Journal of Sociology* 91, 3, 481–510.

Hansen, K.T. and Vaa, M. (2004) *Reconsidering Informality: Perspectives from Urban Africa* (Uppsala: Nordic Africa Institute).

IPCC (2012) Managing the Risks of Extreme Events and Disasters to Advance Climate Change Adaptation. A Special Report of Working Groups I and II of the Intergovernmental Panel on Climate Change [Field, C.B., V. Barros, T.F. Stocker, D. Qin, D.J. Dokken, K.L. Ebi, M.D. Mastrandrea, K.J. Mach, G.-K. Plattner, S.K. Allen, M. Tignor, and P.M. Midgley (eds)]. Cambridge University Press, Cambridge, UK, and New York, NY, USA, 582.

Jürgens, U. and Donaldson, R. (2012) 'A Review of Literature on Transformation Processes in South African Townships', *Urban Forum* 23, 153–163.

Kritzinger, J.N.J. (1996) 'African Cultural Resources in the Struggle against Mammon: The Challenge of Stokvels to the Mission of the Church', *Mission Studies* 13, 109–129.

Liebow (1966) *Tally's corner* (Boston: Little Brown).

I Lindell, E. (2010) 'Introduction: the Changing Politics of Informality – Collective Organizing, Alliances and Scales of Engagement', in I. Lindell (ed.) *Africa's Informal Workers: Collective Agency, Alliances and Transnational Organizing in Urban Africa* (London: The Nordic Africa Institute) pp. 1–32.

Lukhele, A.K. (1990) *Stokvels in South Africa* (Johannesburg: Amagi Books).

MacLeod, A.K., Tata, P., Kentish, J., Carroll, F., Hunter, E. (1997) 'Anxiety, Depression, and Explanation-based Pessimism for Future Positive and Negative Events', *Clinical Psychology and Psychotherapy* 4, 15–24.

Madzwamuse, M. (2010) *Climate Change Vulnerability and Adaptation Preparedness in South Africa* (Cape Town: Heinrich Boll Stiftung).

Matul, M.J., Michael, C., McCord, C., Phily, C. and Harms, J. (2010) *The Landscape of Microinsurance in Africa*, Microinsurance Paper No. 4 (Geneva: ILO).

Mills, E. (2004) *Insurance as an Adaptation Strategy for Extreme Weather Events in Developing Countries and Economies in Transition.* Lawrence Berkeley National Laboratory Report No. 52220. Berkeley, USAID.

Miraftab, F. (2004) 'Neoliberalism and Casualization of Public Sector Services: The Case of Waste Collection Services in Cape Town, South Africa', *International Journal of Urban and Regional Research* 28, 4, 874–892.

Mitchell, J.T. (2003) 'Prayer in Disaster: Case Study of Christian Clergy', *Natural Hazards Review* 4, 1, 20–26. doi:10.1061/(ASCE)1527–6988.

Mokwena, L. (2009) *Municipal Responses to Climate Change in South Africa, The case of eThekwini, the City of Cape Town, and the City of Johannesburg* (Johannesburg: Centre for Policy Studies).

Mukeibir, P. and Ziervogel, G. (2007) 'Developing a Municipal Adaptation Plan (MAP) for Climate Change: The City of Cape Town', *Environment and Urbanization* 19, 1, 143–158.

Murdoch, J. (1999) 'Between the State and the Market: Can Informal Insurance Patch the Safety Net?', *The World Bank Research Observer* 14, 2, 187–207.

Oldfield, S. (2002) 'Embedded Autonomy and the Challenges of Developmental Local Government', in S. Parnell, E. Pieterse, M. Swilling and D. Woolridge (eds) *Democratising Local Government: The South African Experiment* (Cape Town: University of Cape Town Press).

Ostrom, E. (2010) *A Polycentric Approach for Coping with Climate Change. Background paper to the year 2010 World Development Report.* Policy Research Working paper 5095 (The World Bank: Washington DC).

Patel, Z. (2009) 'Environmental Justice in South Africa: Tools and Trade-offs', *Social Dynamics* 35, 1, 94–110.

Porteous, D. (2003) 'Is Cinderella Finally Coming to the Ball? SA micro finance in broad perspective', www.finmark.org.za.

Risse, T. (2011) 'Governance in Areas of Limited Statehood: Introduction and Overview', in T. Risse (ed.) *Governance without a State? Policies and Politics in Areas of Limited Statehood* (New York: Columbia University Press).

Roberts, B. (1994) 'Informal Economy and Family Strategies', *International Journal of Urban and Regional Research* 18, 1, 6–23.

Robinson, M. and White, G. (1997) 'The Role of Civic Organizations in the Provision of Social Sources: Towards Synergy', *Research for action* 37. UNU World Institute for Development Economics Research (UNU/WIDER), Helsinki.

Roth, J. (1999) *Informal Micro-finance Schemes: The Case of Funeral Insurance in South Africa.* Social Finance Unit Working Paper No. 22. International Labour Office, Geneva: International Labor Organization.

Ruggie, G. (2004) 'How to Marry Civic Politics and Private Governance', in CCoEaL affairs (ed.) *The Impact of Global Corporations on Global Governance* (New York: Carnegie Council on Ethics and International Affairs).

Statistics South Africa[1] (2011) Mid-year Population estimates: Statistical Release P0302. http://www.statssa.gov.za/publications/P0302/P03022011.pdf [accessed 29 October 2012].

Statistics South Africa[2] (2012) Quarterly Labour Force Survey, Quarter 1, 2012: Statistical Release P0211 [accessed 29 October 2012]. http://www.statssa.gov. za/publications/P0211/P02111stQuarter2012.pdf.

Stren, R. and Halfani, M. (2001) 'The Cities of Sub-Saharan Africa: From Dependency to Marginality', in K.T. Hansen and M. Vaa, *Reconsidering Informality: Perspectives from Urban Africa* (Uppsala: Nordiska Afrikainstutet).

Thomson, R.J. and Posel, D.B. (2002) 'The Management of Risk by Burial Societies in South Africa', *South African Actuarial Journal* 2, 83–128.

Tissington, K. (2009) *The Business of Survival: Informal Trading in Inner City Johannesburg*, University of the Witwatersrand: Centre for Applied Legal Studies.

Toman, O.M. (2010) 'Shocks, Income Diversification and Welfare in Developing and Transition Countries'. http:// urn:nbn:de:gbv:8-diss-55687.

Tostensen, A., Tvedten, I. and Vaa, M. (2001) *Associational Life in African Cities: Popular Responses to the Urban Crisis* (Nordic Africa Institute).

United Nations Development Programme (1997) Governance for sustainable human development. UNDP policy document, New York.

Valodia, I. (2001) 'Economic Policy and Women's Informal Work in South Africa', *Development and Change* 32, 871–92.

Verhoef, G. (2001) 'Informal Financial Service Institutions for Survival: African Women and Stokvels in Urban South Africa, 1930–1998', *Enterprise and Society* 2, 259–296.

Verhoef, G. (2002) *Money, Credit and Trust: Voluntary Savings Organizations in South Africa in Historical Perspective.* Paper presented to XIII World Congress of the International Economic History Association, Buenos Aires, Argentina, 22–26 July 2002.

Veysey, S. (2012) Climate change risk now better managed by insurers. Business Insurance, January 8, 2012. http://www.businessinsurance.com/article/20120108/ NEWS07/301089987.

White, C. (1998) 'Democratic Societies? Voluntary Association and Democratic Culture in a South African Township', *Transformation* 36, 1–36.

Wood, B. (2004) '24–29 July 2004 Flood Event: City of Cape Town, South Africa', in B. Bouchard, et al. *Improving Flood Risk Management in Informal settlements in Cape Town.* Unpublished Bachelor of Science Interactive Qualifying project, MA: Worcester Polytechnic Institute.

Wood, J. and Shearing, C. (2007) *Imagining Security* (Cullompton: Willan Publishing).

Ziervogel, G. and Smit, W. (2009) *Learning to Swim: Strengthening Flooding Governance in the City of Cape Town*. Working Paper presented at 2009 Amsterdam Conference on the Human Dimensions of Global Environmental Change 'Earth System Governance: People, Places and the Planet'. Amsterdam, 24 December 2009.

10
Business and Climate Change Governance: Conclusions

Tanja A. Börzel, Ralph Hamann and Nicole Kranz

Introduction

South Africa is not only Africa's largest economy, it is also the continent's biggest polluter, and it ranks among the 15 largest emitters of greenhouse gases worldwide. In February 2012, the South African government announced the introduction of a carbon tax to reduce high levels of harmful greenhouse gas emissions. 'To minimise adverse impacts on industry competitiveness and effectively manage the transition to a low-carbon economy', the budget of 2012/13 proposes a 60 per cent tax-free threshold on annual emissions for key sectors, including energy, mining and manufacturers. All sectors except the electricity industry will also be able to claim an additional 10 per cent in relief.[1]

While South Africa would be one of the first countries in the developing world to enact substantive climate change mitigation policies, the tax would exempt a large swath of emissions until 2020. Moreover, South Africa has developed already ambitious and far-reaching environmental legislation since the mid-1990s, including water, biodiversity and recently air. Its capacity for implementing regulations and securing compliance, however, is rather weak. While legal requirements are comprehensive and demanding, details pertaining to the specific behaviour of firms are often not specified. Overlapping responsibilities between several government departments lead to regulatory confusion, contradicting requirements and implementation gaps. Most importantly, the implementation of regulations is in many cases deficient, since state agencies often lack the capacity to monitor and sanction corporate malpractice effectively, although this varies across the country. Enforcement of a carbon tax is likely to be particularly challenging, since the measurement of GHG emissions is difficult, and the tax entails significant costs, which firms are reluctant to bear. The South African industry has already complained that a carbon tax is likely to hit profits, economic growth and employment, particularly in sectors that are already under pressure due to rising energy and wage costs.

While the carbon tax might help to reduce relative greenhouse gas emissions in the long term, it will not by itself turn South Africa into a low-carbon economy. The concessions and implementation problems mentioned above will constrain the effectiveness of the tax, and they illustrate the limitations to state action in tackling the 'wicked problem' of climate change by itself.[2] This also applies, perhaps even more so, to the domain of adapting to climate change. Effective mitigation and adaptation therefore require the active involvement of business. Yet, the governance literature argues that the main incentive for companies to engage in self-regulation is the credible threat of public regulation. Such a shadow of hierarchy is by definition weak in areas of limited statehood (cf. Risse, 2011; Börzel, 2013). So, why should business become active in fighting climate change?

This book has asked how and why firms contribute to climate change governance. South Africa is an appropriate case to explore this question. As a newly emerging economy it is heavily reliant on coal-fired power plants for the energy required to drive its economic growth and provide the poorer parts of the population with electricity. At the same time, South Africa aspires to take a leading role in the global climate change negotiations. Issues of economic growth and social justice may curb its political willingness to follow up on its ambitions, and these ambitions are further compromised by its weak capacity for enforcing ambitious regulation. Nevertheless, the contributions to this volume find ample evidence for business responses to climate change, both with regard to mitigation and adaptation. We contrast our findings on South Africa with evidence from Kenya and Germany in order to elicit different patterns due to variation in the degree of statehood as well as other framework conditions. Thus, while this book mainly focuses on South African experiences, it allows for insights into the role of business in climate change governance in other contexts.

Business responses to climate change

Our empirical chapters are grouped in terms of an emphasis on either mitigation or adaptation, although we also recognise the linkages and overlaps between them. In addition, in our Introduction (Chapter 1) we suggest a mapping of business responses with regard to the predominant locus of intent and effort, on the one hand, and the extent of effort and the time-scale applied to calculating investment returns, on the other.[3]

The empirical chapters cover the various options created by this map. For a start, Farai Kapfudzaruwa (Chapter 2) finds that listed companies in South Africa and Kenya represent a range of strategic postures in response to climate change, as evidenced in their public reports. His four clusters of firms occupy varying spaces on the map in Figure 1.4 in Chapter 1. The laggards display a very limited response and if anything adopt 'low-hanging fruit' approaches with an internal focus and low asset specificity. The majority of Kenyan (84 per cent) and South African (61 per cent) firms are in this category. Emergent planners expand their tentative efforts in both dimensions, including some engagement externally and at least planning investments without immediate returns. Seventeen per cent and 7 per cent of the sample firms in South Africa and Kenya, respectively, are in this group. The efficiency drivers are characterised by relatively greater investments, especially within the firm, often with a focus on energy reductions (represented by about 10 per cent of companies in both countries). Finally, the visionaries have a broad array of initiatives that also include some investments in systemic change beyond the firm's boundaries. There are no such companies in Kenya in this cluster (though Unilever represents many of these traits in that country), while 12 per cent of South African companies are in this group (see Figure 10.1 for a schematic illustration of these clusters in our 'how' map).

While Kapfudzaruwa focuses on corporate sectors to explore variations in corporate responses to climate change, Christian Thauer (Chapter 3) shows that there are also significant variations within sectors. He looks at the adoption of environmental process standards as one of the specific responses identified in our Introduction and finds differences in this adoption between four multinational automotive companies with manufacturing facilities in South Africa. Not only are there differences between the firms, there are also temporal differences within individual firms. Furthermore, these temporal differences are not always improvements – one of the firms' process performance actually declined in recent years.

While most of the chapters in this book adopt an organisational level of analysis, John Fay's chapter (Chapter 4) focuses on renewable energy projects and their financing. The renewable energy industry is a prominent example of ongoing efforts to make climate change mitigation, and latterly also adaptation, a feasible proposition for private sector investment. Risk-adjusted return on investment, and hence the expected time-scale for such returns, play a prominent role in investors'

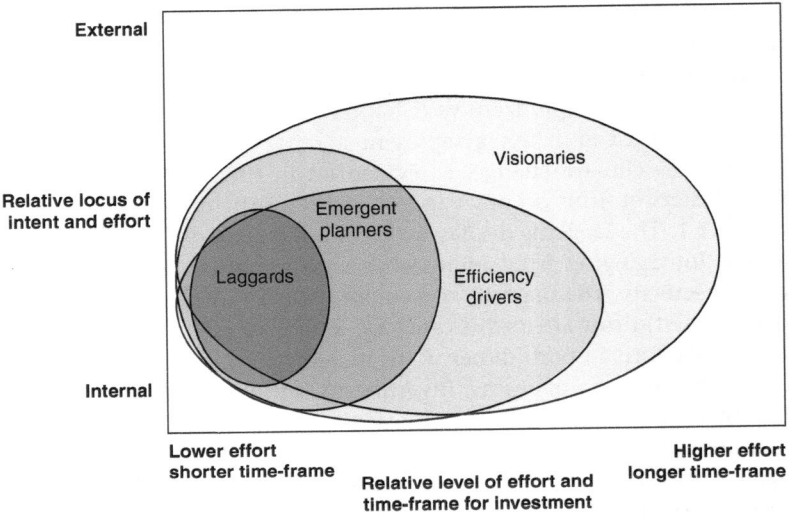

Figure 10.1   Schematic illustration of Kapfudzaruwa's company clusters' position in the locus/time-frame map of business responses to climate change (as described in Chapter 1)

decision-making. There is hence a zone in the bottom left quadrant in Figure 1.4 of Chapter 1, which delineates projects that are more likely to receive bank loans or equity investment. Crucially, Fay's chapter argues that this zone is diminished in areas of limited statehood by investors' cost of capital calculations (see Figure 10.2 for an illustration).

Christopher Kaan and Stine Klapper (Chapter 5) expand the scale of analysis to the external environment of firms, focusing on the drivers for their participation in a collective initiative to enhance energy efficiency in South Africa. It is thus not the firms' commitment to energy efficiency per se, that is of primary interest, but their involvement in a collective initiative. The point of departure for their analysis is that there have been numerous companies making substantial energy reduction commitments as part of the Energy Efficiency Accord launched by the National Business Initiative (NBI) in 2005. This voluntary agreement, developed in cooperation with the South African government, sets targets for a decrease in energy consumption and establishes reporting and information exchange mechanisms to ensure implementation. The standards go well beyond what most companies would have committed to at an individual level.

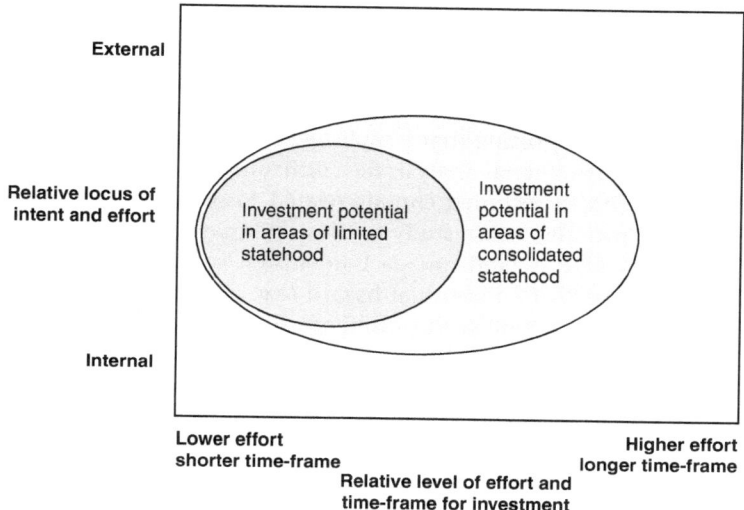

Figure 10.2 Schematic illustration of John Fay's analysis of renewable energy projects' financing prospects as a function of statehood

Nicole Kranz and Nadine Methner (Chapters 6 and 7 respectively) start the book's discussion of business contributions to climate change adaptation, focusing in particular on water catchment management. Kranz finds that South African companies mainly engage in coping activities and relatively conventional strategies to reduce the risks of climate change, for instance increasing the efficiency of internal water use, recovering and recycling wastewater or securing additional access rights to water resources. Some companies engage in more substantial adjustments and emerging transformative approaches, for example by contributing to proactive catchment management and the development of water infrastructure in collaboration with public authorities, or by assisting their suppliers in the efficient use of natural resources. Methner analyses one of the few cases of a company in South Africa engaging in system transformation. Woolworth's Farming for the Future programme is a prominent example of a proactive water management strategy that aims at ensuring sustainable water supply by systematically involving the supply chain and seeking to change farming practices. German firms, by contrast, are more concerned with mitigation. Their adaptation activities focus on coping, and they are relatively less active than their South African counterparts.

The emphasis on corporate efforts to influence the broader socio-ecological and governance context in responding to climate change is also prominent in Chapter 8, by Tom Herbstein, Jan Froestad, Deon Nel and Clifford Shearing, on the insurance industry. Some insurance companies are recognising that established risk-management technologies that rely on actuarial analysis face difficulties in responding to the exponential rise in recorded climate-related losses (or 'climate risks'). The chapter includes a case study to illustrate that social influences on a landscape (such as land-surface hardening) can contribute to over half of the impact of an eventual hazard (e.g. a flood). This creates an opportunity for the industry to influence these social and governance drivers of climate risk, but there are a number of constraints to such a proactive approach.

Finally, Moliehi Shale (Chapter 9) also investigates responses to climate risks, but she focuses on survivalist, informal enterprises in poor urban communities affected by flooding. She shows that such businesses largely eschew formal insurance in their response to flooding or indeed other hazards. Rather, they rely on religious and cultural networks developed through burial societies to manage risks. This provides a profound challenge to public policy and private sector practitioners, as well as to academic researchers, who commonly have little understanding of or access to such religious and cultural networks.

Drivers of business responses to climate change

The Introduction (Chapter 1) identifies three sets of drivers that may motivate business organisations to contribute to mitigation and adaptation. Rather than providing competing explanations, they combine into different configurations that account for the business responses we have identified.

With regard to mitigation, we find that business responses within the firm aimed at curbing negative externalities of their production activities, such as reducing GHG emissions, are indeed driven by a shadow of hierarchy. The overall better performance of South African firms, as compared to Kenyan ones, in responding to climate change, is largely explained by the more limited statehood of Kenya, which has not sufficient regulatory capacity to enact and enforce climate change regulation (Kapfudzaruwa, Chapter 2).

At the same time, the absence or ineffectiveness of state regulation can also induce companies to engage in climate change governance. Companies have little faith in the capacity of the South African state to ensure reliable energy supply. The shadow of anarchy has been a major reason for many of them to join the Energy Efficiency Accord (Kaan and Klapper, Chapter 5). Thus, the shadows of hierarchy and anarchy can work side by side, affecting different aspects of a governance problem. The shadow of anarchy appears to be more relevant for climate change adaption, however, given its greater task complexity (as discussed in more detail below).

Statehood and the lack thereof do not only drive companies to engage in climate change mitigation in order to anticipate (more) state regulation and reduce risk and uncertainty regarding the provision of collective goods and services such as energy security or clean water. These institutional drivers also have an impact on market incentives for renewable energy development. The comparison between Germany and South Africa clearly demonstrates that limited statehood increases the costs of capital for renewable energy projects, which in turn influences the required tariff and price differentials between electricity from renewable sources and fossil fuels (Fay Chapter 4). In other words, to be effective, market incentives may require a shadow of hierarchy, which calls into question the extent the former can substitute for the latter.

Yet there are economic incentives for business to engage in climate change governance that are independent from statehood and that manifest at the organisational level. Firms voluntarily introduce process regulation for enhanced water and energy efficiency to secure substantial investments in production facilities and technology with a long pay-off period (asset specificity; Thauer Chapter 3). By helping head offices of multinational firms maintain organisational control and authority and reduce the vulnerability to the uncertain and insecure investment environment, voluntary standards such as ISO 14001 substantially contribute to climate change mitigation. Asset specificity is an organisational driver that has not figured prominently in the literature on business and governance (but see Thauer, 2013; Thauer, forthcoming 2014). It clearly deserves more attention, since areas of limited statehood entail substantial uncertainties for companies (Börzel and Thauer, 2013).

Another organisational driver relates to corporate culture and leadership, particularly when investments in climate change mitigation or adaptation do not have a clear, short-term return. Hence, a distinguishing feature of firms in Kapfudzaruwa's visionaries cluster is an explicit leadership commitment to sustainable development and/or climate change mitigation. This is also apparent in Methner's case study of Woolworths – an explicit corporate strategy on sustainable development, coupled with corresponding performance indicators, created a culture that was conducive for the Farming for the Future initiative to be proposed and implemented (see also Hamann, Methner and Nilsson, 2012).

While the two shadows and asset specificity explain why business engages in climate change mitigation in the first place, brand name, market orientation, task complexity and problem pressure appear to be major drivers for firms to extend their responses beyond the purview of their production sites and to engage with other actors in the fight against climate change. Joining the Energy Efficiency Accord, a voluntary collective agreement initiated by the National Business Initiative, has been motivated largely by the perceived incapacity of the South African state to provide for energy security, on the one hand, and the threat of a carbon tax on the other. Yet some companies have been more responsive to the combined effect of the shadows of anarchy and hierarchy than others. This is explained by the reputational gains companies hope to reap by signalling their commitment to climate change mitigation. Some companies are more sensitive than others about their image, because they have a brand name to protect, they export to high-regulating countries where consumers care about the carbon footprint of the products they purchase, and they are subject to NGO or community pressure (Kaan and Klapper, Chapter 5; Kapfudzaruwa, Chapter 2).

Whereas reputational concerns relate to brand name and market orientation and, hence, are organisational drivers, task complexity and problem pressure are issue-area specific. Reducing GHG emissions as such does not require companies to cooperate with others. Energy security or water management, however, are complex tasks that resemble common-pool resource management (Ostrom, 1990) and can hardly be addressed by in-house policies only (Kranz, Chapter 6; Methner, Chapter 7). The need to engage in a collective response is increased for energy-intensive companies or sectors that face greater problem pressure (Kranz, Chapter 6; Methner, Chapter 7; Kaan and Klapper, Chapter 5; Kapfudzaruwa, Chapter 2).

For adaptation, it is again a combination of institutional and issue area-specific drivers that motivates firms to move beyond coping, to engage in more substantial adjustments and system transformation. If the state has only limited capacity to address complex tasks, such as securing efficient energy or water supply, companies seek to engage others in their adaptation activities. This is particularly the case when the combined effect of the shadow of anarchy and task complexity is magnified by problem pressure, for example because common-pool resources such as clean water are becoming depleted (Methner, Chapter 7; Kranz, Chapter 6).

The motivation of companies to engage in broader, longer-term activities is reinforced by organisational drivers, such as brand name, market orientation and knowledge and learning (Methner, Chapter 7; Kaan and Klapper, Chapter 5). Voluntary agreements, multi-stakeholder forums or business associations do not only signal commitment to climate change governance to governments, consumers and shareholders. They can also help strengthen the capacities of firms and foster learning processes (Methner, Chapter 7; Kaan and Klapper, Chapter 5; Kapfudzaruwa, Chapter 2). A corporate culture of, and capacity for, organisational learning appears to be especially pertinent in the adaptation to the risks of climate change, particularly since in areas of limited statehood state regulation is unlikely to regulate the risks (Herbstein, Froestad et al., Chapter 8 and Methner, Chapter 7).

While the findings on adaptation largely correspond to the dynamics that we find with regard to mitigation, two stand out in particular. First, the shadow of hierarchy can delay rather than induce company responses to climate change. It is precisely the expectation that the German government will prescribe adaptive policies that prevents companies in that country from taking action (Kranz, Chapter 6). Likewise, state regulation makes insurance companies refrain from exploring new technologies to shape social drivers of climate risk rather than merely perfecting assessment tools to exclude uninsurable risk (Herbstein, Froestad et al., Chapter 8). Thus (the anticipation of) state regulation is not always conducive to climate change responses by companies, particularly when the task is complex, rendering business responses more costly, and problem pressure is low (Kranz, Chapter 6).

Second, informal networks based on shared religious beliefs and social norms provide an alternative form of collective responses to climate change. Developing countries and emerging economies feature informal settlements where small informal businesses dominate the low-end markets. Despite efforts by insurance companies, small-business owners prefer their own social networks over formal insurance to respond to climate change related hazards, such as flooding (Shale, Chapter 9). While insurance policies are affordable and would compensate for financial losses, burial societies are present in the community and provide spiritual and social rather than solely economic support in case of loss.

The contributions to this volume confirm the relevance of the various drivers we have identified in the Introduction. Rather than concluding that everything matters, our findings allow us to specify what exactly the different drivers explain and how they may relate to each other.

The shadows of hierarchy and anarchy do the most in explaining cross-country variation. While regulatory capacity varies significantly between South Africa, Kenya and Germany, the case of energy security shows that level of statehood may vary even across issues within a particular policy area (Kaan and Klapper, Chapter 5). The South African government is in a better position to mandate the reduction of GHG emissions than to provide energy security, which is a more complex task. Likewise, mitigation generally represents a lower level of task complexity than adaptation.

Cross-sectoral differences in business responses to climate change are best accounted for by problem pressure, on the one hand, and sectoral institutions, such as business associations, on the other. If we control for other drivers at the country and sector level, asset specificity, brand name, market orientation, and NGO activism largely explain why some companies are more responsive than others (Thauer, Chapter 3; Kaan and Klapper, Chapter 5; Kapfudzaruwa, Chapter 3; Kranz, Chapter 6; Methner, Chapter 7). Moreover, particularly with regard to adaptation, leading companies are characterised by strong leadership and a corporate culture conducive to organisational learning. Finally, reputation concerns driven by brand name, market orientation and NGO activism, as well as task complexity and problem pressure are especially pertinent in motivating companies to engage in collective responses, rather than confining their activities to in-house programmes or supply-chain regulation.

The findings of this volume thus suggest a multilevel model, in which institutional, organisational and problem-specific factors interact to explain variations at the country, sector and company level. While most contributions focus on one particular level, the cluster analysis of Kapfudzaruwa exemplifies the explanatory power of such a 'funnelling' approach. We can also show that the various drivers are alternative, rather than competing in inducing business responses to climate change. Not only do they often work simultaneously, certain drivers reinforce each other or mitigate their effects. We identify five interaction effects (see Figure 10.3). Task complexity intensifies the shadow of anarchy (1) but weakens the shadow of hierarchy (2). The reinforcing effect on the shadow of anarchy is even more likely if problem pressure is high (1). On the other hand, low problem pressure further mitigates the shadow of hierarchy (2). Thus, energy-intensive companies are particularly vulnerable to the failure of the South African state to provide the complex task of energy security. Energy shortage immediately threatens their production base (Kaan and Klapper, Chapter 5). The same applies to companies that rely on sufficient (clean) water for their production processes. German companies, by contrast, are less exposed

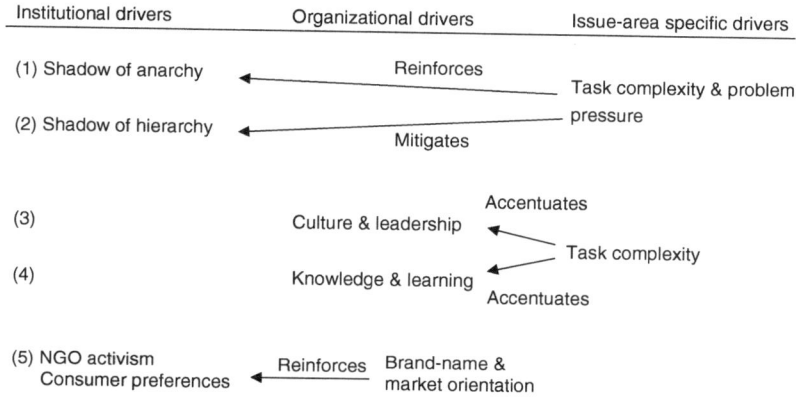

Figure 10.3 Three interaction effects between institutional, organisational and issue-area specific drivers

to climate change risks and thus do not see the need to anticipate state regulation (Kranz, Chapter 6). We find a similar interaction effect between task complexity, on the one hand, and corporate leadership and organisational learning on the other. The more complex the task, the more likely not only the need for business to compensate for the failure of the state to accomplish it (1) but also the more important are corporate leadership (3) and a culture of learning (4). Finally, characteristics such as a brand name and high-end market orientation do not just increase reputational concerns; firms that have an image to protect are much more vulnerable to NGO and local community pressure, particularly if they cater to high-end markets where consumers are more likely to care about climate change (5).

Business and governance in areas of limited statehood

South Africa provides a useful case study for our investigation into the role of business in climate change governance, because it has a relatively sophisticated economy – with well-established finance and retail sectors complementing the historically dominant mining sector – while at the same time exhibiting various aspects of limited statehood. Furthermore, given its reliance on coal-fired energy generation and historically cheap electricity, it faces significant challenges in reducing its GHG emissions, and it also runs into difficulties as a result of increasing temperatures and changes in water availability. In other words, companies operating in South Africa face significant pressures related to both mitigation and adaptation; and although the state is making ambitious plans in the mitigation domain in particular, there are limitations to its capacity to enact and enforce such regulations.

These circumstances, in combination with the comparative analyses in the chapters by Kapfudzaruwa, Fay and Kranz, allow for the extrapolation of our findings to a range of other settings. This is particularly so with regard to the interaction effects between the drivers for business responses, as outlined above. We argue that the shadow of hierarchy has an important role in motivating business efforts, especially when the complexity of the task is not very significant and thus (the threat of) state regulation is more compelling. Carbon tax is an example of this. The influence of the shadow of hierarchy is dependent on state capacity; as we move to countries where statehood is more limited than in South Africa, such as Kenya, it declines, and we see less committed responses by companies.

At the same time, as the complexity of the task increases – as in the case of ensuring energy or water supply – the capacity of the state to enforce rules or provide public services diminishes, and thus the shadow of hierarchy plays a relatively lesser role, at least in areas of limited statehood. Instead, the shadow of anarchy becomes more influential, as some companies attempt to innovate to ensure, for instance, long-term reliability in the supply of products or mitigation of climate risks to clients. In areas of more consolidated statehood, the role of the shadow of anarchy is less prominent, and Kranz's analysis of adaptation by companies in Germany suggests that the assumption that the state will become active in regulating climate change may even delay private sector action. In areas of weaker statehood than that of South Africa, it is not clear whether the shadow of anarchy motivates innovation in the private sector. Even in South Africa, such innovation was attempted by relatively few companies (such as Woolworths and Santam) and it stands to reason that organisational drivers (committed leadership and organisational learning) are required to give rise to such organisational responses to the shadow of anarchy in the arena of climate change issues. This is a question for further research.

Our findings have a number of important implications for research on climate change and the more general discussions on governance and management. We have supported the view that the complexity of climate change mitigation and adaptation creates inherent constraints to the ability of states to address these problems unilaterally, and that this gives rise to an important role for businesses and other non-state actors to make proactive contributions (Brunner and Lynch, 2010). We have shown that this is particularly so in areas of limited statehood. However, this is not a matter of either 'top-down' state action or 'bottom-up' business responses.

Instead, we have found empirical evidence of a complex array of interactions between the presence or absence of both state regulation and private sector action. But these interactions are more varied, less ambitious and probably less effective than the 'virtuous cycle of push and pull between public and private sectors' called for by Christiana Figueres.[4]

Our contribution to the management literature rests, first, on the explicit focus on the role of statehood. The state has been relatively absent in many discussions on corporate social or ecological responsiveness (Matten and Moon, 2008) or even cross-sector collaboration (Crane, 2010). Much of the institutional theory literature in the management field assumes that if certain regulatory institutions are formally in place, these will create binding constraints on organisational action (e.g. Hoffman, 1999). This conclusion is likely to be related to the empirical focus of the management literature on North American and European countries.[5] Our analysis recognises that the existence of formal rules does not ensure their enforcement, and so business responses to state regulations will vary depending on the level of statehood. This limited role for formal institutions is particularly pronounced in low-income areas, where businesses are enmeshed in informal institutions and networks that can be more important than formal rules in guiding behaviour (Börzel and Hönke, 2012; Börzel, Hönke and Thauer, 2012). This is illustrated vividly by Shale's chapter on how small, survivalist businesses prefer the social and spiritual benefits provided by burial societies to the products offered by the formal insurance companies. We thus show that the broader challenges posed by informality to management scholarship (Bruton et al., 2012) apply in the domain of climate change in particular ways.

Second, business responses to climate change and climate change adaptation, in particular, represent a relatively under-researched area in the management literature, associated with an inability to conceptualise the possible impact on business of 'massive discontinuous change' (Winn et al., 2011). Furthermore, perhaps because of the above-mentioned bias toward areas of consolidated statehood, the business responses that are commonly expected or analysed are relatively reactive, and are focused within the boundaries of the firm or its immediate environment, even in industries that are likely to be severely affected by climate change, such as the ski industry (Tashman, 2011). Although the potential for system transformation has been theorised by Moser and Ekstrom (2010), there does not seem to be much empirical evidence of this. Our analysis, however, has shown that in areas of limited statehood at least some companies attempt system transformation and devote significant resources to engaging the state and other role-players to this effect. The shadow of anarchy is an important motive for this, though additional, organisational drivers play a role: the companies illustrating this approach have brand names and asset-specific investments to protect, as well as an explicit, public leadership commitment to sustainable development. More research is called for to understand better the antecedents of such transformational efforts among companies.

Finally, the multifaceted, complex nature of the climate change domain allowed us to develop a model of interactions between institutional, organisational and issue-specific drivers of business responses to limited statehood that travels to other issue areas, and hence, offers important contributions to the broader governance literature. In particular, we highlight the important role of task complexity, which has arguably been underplayed in much of the 'new' governance literature (except perhaps when discussing implications for performance management: Perry et al., 2006). A particularly salient implication is that varying levels of task complexity give rise to different roles for shadow of hierarchy and shadow of anarchy – key constructs in the literature on limited statehood (see, e.g., Börzel and Risse, 2010). Thus, we argue that the shadow of hierarchy plays a relatively greater role in climate change mitigation, because the state can exert a more credible shadow of hierarchy through instruments such as a carbon tax in this domain. Meanwhile, the shadow of anarchy is particularly prominent in climate change adaptation, because of the higher task complexity in this domain.

Furthermore, the discussion of governance in areas of limited statehood has generally assumed that companies are exposed to either the shadow of hierarchy or the shadow of anarchy (Börzel and Risse, 2010). The analyses in this book lead us to argue that these two seemingly opposed drivers can be present at the same time, depending on the task complexity of different aspects of the overarching problem. For instance, companies joined the Energy Efficiency Accord because they worried both about the state raising energy costs or imposing supply constraints (shadow of hierarchy) and about the state's inability to ensure long-term energy supply (shadow of anarchy) (Kaan and Klapper, Chapter 5).

Finally, market incentives are important drivers for business contributions to (climate change) governance in areas of limited statehood (Börzel and Thauer, 2013). The chapter by Fay, however, highlights that effective state regulation influences the costs of capital, which raises doubts whether economic instruments can substitute for the absence of a credible shadow of hierarchy in inducing companies to get involved in reducing their negative externalities or directly providing collective goods.

Lessons for climate change governance

The findings presented in this book have implications for practitioners and policy-makers. They confirm that business organisations are important actors in climate change mitigation and adaptation, that there is significant variation in their responses, and that these responses are motivated by a range of interrelated factors at various levels. We thus agree that business organisations can be important actors that need to be proactively engaged in addressing complex challenges such as climate change (Pinkse and Kolk, 2010; Pattberg, 2010). This is particularly so in areas of limited statehood, where national and local government agencies are overwhelmed with other priorities and struggle to enforce commonly binding rules.

However, we also argue that the state matters. Suggesting that some companies proactively engage in addressing collective problems when the state does not do so in no way translates into reducing an emphasis on the need to build the state's capabilities. When comparing South Africa and Kenya, or different policy arenas within South Africa, our findings clearly stress the important role of the state's 'shadow of hierarchy' in motivating corporate responses to climate change. Our findings also suggest an important role for state institutions in market-based incentive instruments, that is, initiatives that have been recommended, at least in part, as an alternative or even a replacement for state intervention. The implication is that market-based instruments should not be seen as a mechanism to compensate for limited state capacity, and the latter needs to be included in efforts to develop the former.

So we need to temper the view that effective responses to climate change can be based primarily on 'bottom up' initiatives by civil society and business. We do find innovative initiatives among business organisations important, and these are likely to have significant demonstration effects beyond their more immediate impacts. But they are not very common. Stronger state action is required to raise the bar across companies that do not face sector- or company-specific pressures. We thus reinforce the argument that nation states are at the centre of decision-making in climate change governance, as they have the power to issue and enforce domestic regulation, and to allocate financial resources and responsibilities (Meadowcroft, 2010; Newell and Bulkeley, 2010).

But this is not only about enhancing states' capabilities for making and enforcing rules, and thus a re-emphasis on the well-worn call for building state capacity in 'weak governance' zones. It is also about the kind of rule making and enforcement that is required, especially in policy areas characterised by high task complexity, such as climate change adaptation. Perhaps paradoxically, we have found that the state's ability to engage in an adaptive, participatory manner in such policy domains (as recommended, for instance, by Berkhout, Hertin and Arnell, 2004; Loorbach et al., 2009) may be hampered in areas with strong, consolidated states. The German state, for instance, may struggle to move to such adaptive responsiveness from its established emphasis on hierarchy and rule enforcement. German companies, meanwhile, are less likely to become proactively involved in climate change adaptation, because of their expectation that the state will 'deal with it'. So the building of state capacity needs to take place across the spectrum of statehood levels, and it needs to respond to the different kinds of state action that are required in response to different levels of task complexity.

While our first overarching conclusion for policy-makers is that the state still matters, our findings also suggest that there are a number of factors beyond the state that influence business organisations' responses to climate change. One of the implications is that these various factors and their interrelationships will need to be borne in mind by policymakers, lest a single-minded focus on the state gives rise to unintended consequences. It also means that there is a broader constellation of actors that can be, and probably need to be, proactively engaged in developing a systemic response to climate change, including business managers, educators and trainers, auditors, civil society activists and customers. Rather than a binary relationship between the state and companies, the image that emerges is one of a web of interrelationships, where the opportunity for agency is also more widely distributed. This includes a broader array of opportunities for the state as well, in influencing the way in which, for instance, finance institutions, development aid agencies or consumer protection bodies include climate change considerations in their interactions with companies.

Management educators and consultants will need to emphasise, in their discussions with managers, that there are good reasons for companies to adopt strategic responses to climate change, to implement sound environmental management programmes, to innovatively maintain or create public goods where the state is unable to do so, and to engage in collective and collaborative initiatives to address complex problems. Our findings identified leadership- and organisational-culture as playing an important role in driving firms' climate change responses. We thus confirm arguments in an emerging literature on the important role of leadership in the broader field of corporate responsibility, and the significant capabilities required of effective sustainability leaders in sensing and responding to complex problems (Metcalf and Benn, 2013). More research is needed, however, to understand better whether and how such leadership abilities and proclivities can be fostered. In the meantime, we suggest that climate change issues require more attention in a diversity of training and education settings, including MBA courses and executive management education, in particular.

We suggest a particularly important role for innovative corporate initiatives such as those by Woolworths and Santam discussed in this book (see Chapters 7 and 8 by Methner and Herbstein and colleagues, respectively). Their impact goes well beyond the immediate outputs of the initiative itself. They have a vital potential to illustrate the different ways of understanding a complex problem such as climate change, and of responding in a committed and novel manner. Woolworths' Farming for the Future, for instance, is inviting replication by competitors and changing the way retailers and farmers interact on climate change related issues. Such innovative initiatives are changing the way in which the above-mentioned web of interrelationships is configured. Over and above encouragement through awards and the like, the implication is that there needs to be a greater appetite for private- and public-sector managers to engage in such experimentation. These capabilities are a crucial dimension of the above-mentioned imperative for the building of state capacity (see Hamann and April, 2013).

Finally, our findings have emphasised the importance of collective business initiatives and cross-sector collaboration in responding to climate change. This is particularly so for those issues characterised by high degrees of task complexity, which in turn give rise to relatively stronger shadow of anarchy (e.g. ensuring reliable energy or water supply). Under these conditions, collective efforts are conducive to more systemic impacts than individual action. But the state still plays an important role in providing and enforcing the ground-rules for such engagements (Hamann, 2013). The ability to encourage, convene or facilitate collective and collaborative initiatives is yet another crucial element in the state capacity building effort.

Notes

1. http://af.reuters.com/article/southAfricaNews/idAFL5E8DM4VA20120222, last access January 3, 2013. Note that the proposed, broadly applicable carbon tax follows an earlier regulation that levies carbon taxes on new passenger vehicles.
2. The efficacy of carbon taxes in reducing GHG emissions is not guaranteed, as illustrated in the experience of developed economies that have imposed such

a tax. For instance, while Sweden's GHG emissions have diminished since the imposition of such a tax, Norway's have continued to increase.
3. Note that the latter dimension is usefully discussed in terms of asset specificity as an explanatory variable in chapter 3, but in mapping corporate responses we use it descriptively.
4 http://www.cbi.org.uk/media/1044924/green_business_dinner_-_christiana_figueres.pdf, last accessed January 24, 2013.
5 This focus is particularly pronounced in top-tier management journals, as found in a bibliometric analysis of these journals (Hamann, 2012).

References

Berkhout, F., Hertin, J. and Arnell, N. (2004) Business and Climate Change: Measuring and Enhancing Adaptive Capacity, in *Tyndall Centre Technical Report*, edited by T. A. project: Tyndall Centre for Climate Change Research.

Börzel, T.A. 2013. 'Business and Governance in Areas of Limited Statehood', in T.A. Börzel and C.R. Thauer (eds) *Business and Governance in South Africa. Racing to the Top?* (London: Palgrave Macmillan).

Börzel, T.A. and Hönke, J. (2012) *Security and Human Rights. Mining Companies between International Commitment and Corporate Practice* (München: Schriftenreihe des Centrum für Governance Forschung).

Börzel, T.A., Hönke, J. and Thauer, C. (2012) 'How Much State Does it Take? Corporate Responsibility, Multinational Corporations, and Limited Statehood in South Africa', *Business and Politics* 14, 3, 1–34.

Börzel, T.A. and Risse, T. (2010) 'Governance without a state: Can it work?' *Regulation & Governance* 4, 2, 113–134.

Börzel, T.A. and Thauer, C. (eds) (2013) *Business and Governance in South Africa. Racing to the Top?* (Houndmills: Palgrave Macmillan).

Bruton, G., Ireland, R. and Ketchen, D. (2012) 'Toward a research agenda on the informal economy', *Academy of Management Perspectives* 26, 3, 1–11.

Crane, A. (2010) 'From governance to Governance: On blurring boundaries', *Journal of Business Ethics* 94, 17–19.

Hamann, R. (2013) 'Cross-sector Partnerships in Areas of Limited Statehood', in M.M. Seitanidi, and A. Crane (eds), *Social Partnerships and Responsible Business: A Research Handbook* (London: Routledge).

Hamann, R. and April, K. (2013) 'On the Role and Capabilities of Collaborative Intermediary Organisations in Urban Sustainability Transitions', *Journal of Cleaner Production* 50, 12–21.

Hamann, R., Methner, N. and Nilsson, W. (2012) 'The Evolution of a Sustainability Leader: The development of strategic and boundary spanning organizational innovation capabilities in Woolworths', in *EURAM 12 conference*, Rotterdam.

Hoffman, A.J. (1999) 'Institutional evolution and change: environmentalism and the U.S. chemical industry', *Academy of Management Journal* 42, 4, 351–371.

Loorbach, D., Van Bakel, J.C., Whiteman, G. and Rotmans, J. (2009) 'Business Strategies for Transitions Towards Sustainable Systems', *Business Strategy and the Environment* 19, 2, 133–146.

Matten, D. and Moon, J. (2008) '"Implicit" and "explicit" CSR: A conceptual framework for a comparative understanding of corporate social responsibility', *Academy of Management Review* 33, 2, 404–424.

Meadowcroft, J. (2010) 'Climate Change Governance', *Background Paper to the 2010 World Development Report* WPS4941.

Metcalf, L. and Benn, S. (2013) 'Leadership for Sustainability: An Evolution of Leadership Ability', *Journal of Business Ethics* 112, 3, 369–384.

Moser, S.C. and Ekstrom, J.A. (2010) 'A framework to diagnose barriers to climate change adaptation', *Proceedings of the National Academy of Sciences (PNAS)* 107, 51, 22026–22031.

Newell, P. and Bulkeley, H. (2010) *Governing Climate Change* (London/New York: Routledge).

Ostrom, E. (1990) *Governing the Commons. The Evolution of Institutions for Collective Action* (Cambridge: Cambridge University Press).

Pattberg, P. (2010) 'Public-private Partnerships in Global Climate Governance', *Wiley Interdisciplinary Reviews: Climate Change* 1, 2, 279–287.

Perry, J., Mesh, D. and Paarlberg, L. (2006) 'Motivating employees in a new governance era: The performance paradigm revisited', *Public Administration Review* 66, 4, 505–514.

Pinkse, J. and Kolk, A. (2010) 'Challenges and Trade-offs in Corporate Innovation for Climate Change', *Business Strategy and the Environment* 19, 4, 261–272.

Risse, T. (2011) 'Governance in Areas of Limited Statehood: Introduction and Overview', in T. Risse (ed.) *Governance without a State? Policies and Politics in Areas of Limited Statehood* (New York: Columbia University Press).

Tashman, P. (2011) 'Climate change and environmental performance: A longitudinal study in the US ski industry', *Academy of Management Proceedings* 1, 1–6.

Thauer, C.R. (2013) 'Goodness Comes From Within. Intra-organizational Dynamics of Corporate Social Responsibility', *Business and Society, Online First*: http://intl-bas.sagepub.com/content/early/2013/04/17/0007650313475770.full.pdf+html (24th April 2013).

Thauer, C.R. (forthcoming 2014) *Internal Drivers of Corporate Social Responsibility. Managerial Dilemmas and the Spread of Global Standards* (Cambridge: Cambridge University Press).

Winn, M., Kirchgeorg, M., Griffiths, A., Linnenluecke, M. and Günther, E. (2011) 'Impacts from climate change on organizations: a conceptual foundation', *Business Strategy and the Environment* 20, 3, 157–173.

Index